AXIS AND CIRCUMFERENCE

STEPHEN A. WAINWRIGHT

AXIS AND CIRCUMFERENCE

The Cylindrical Shape of Plants and Animals

HARVARD UNIVERSITY PRESS
Cambridge, Massachusetts
London, England
1988

Library of Congress Cataloging in Publication Data
Wainwright, Stephen A., 1931–
    Axis and circumference.

    Bibliography: p.
    Includes index.
    1. Morphology.   I. Title.   II. Title: The cylindrical shape of
plants and animals.
QH351.W25   1988   574.4   87-21099
ISBN 0-674-05700-7 (alk. paper)

Preface

After creating a new genre of biomorphic sculptures from hollow cylinders, Frank Smullin said, "I have found an intriguing duality: contrast between axis and circumference can be exploited for expressive purposes" (Fane and Smullin, 1984). Smullin's observation caused me to see the external shape and internal structure of plants and animals as an expression—achieved via evolution—of the same duality of the cylindrical body's axis and circumference.

Normally we use information about features at lower levels in the structural hierarchy to explain phenomena at higher levels. For example, the rhythmic contraction of cardiac muscle cells is the mechanism that produces the heartbeat. The intent of this book is to turn this process around. Is there a morphological feature of whole organisms that could explain and justify what we already know about design features at lower levels? If so, such a feature of the whole organism might lead us directly to other features and other connections among features of which we are now ignorant.

The idea that science might be done in this upside-down direction was given to me by Deborah Gordon and elaborated for me by Rupert Riedl and Terri Williams. Their thoughts played among exciting thoughts of a few other people and led me to the synthesis presented in this book. I want to introduce you to these sources: Agnes Arber (1954, reprint 1985), Jacob Bronowski (1978), R. B. Clark (1964), Buckminster Fuller (Marks, 1960), J. E. Gordon (1968, 1978),

Laurence Picken (1960), Rupert Riedl (1978, 1984), E. S. Russell (1916, reprint 1982), Cyril S. Smith (1981), Knut Schmidt-Nielsen (1972), Steven Vogel (1981), and the father of all this kind of thought, D'Arcy Thompson (1917, précis by Bonner, 1968). To pick up any of these books is to commit your time and imagination until you finish reading it.

I dedicate this book with profound gratitude to my students, my colleagues, and my teachers.

Contents

Structure without function is a corpse and function without structure is a ghost.

Vogel and Wainwright, 1969

1
THE IMPORTANCE
OF SHAPE

All animals, including hu-
mans, pay attention to other
organisms in order to ensure appropriate behavioral response
to them. But human responses to organisms go beyond those
prompted simply by fear and hunger, the responses that seem
to control animal behavior. We can even be fascinated by the
endless variety of plants and animals. Indeed, much creative
thought has gone into musings on the shape of plants and
animals over the ages. Concepts, theories, poems, essays,
songs, and treatises of philosophy, religion, engineering, history, art, and science have sought the meaning of organismic
form.

Visual information dominates our perceptions of plants and
animals: we name and classify organisms first by their shapes
and colors. Later we want to understand what we see. We
want to know whether there is some single observation about
the shapes of all plants and animals that makes sense, a statement that can explain how all organisms are different from
nonliving forms and yet show that organisms belong on
Earth's surface along with rocks, rivers, oceans, mountains,
and the atmosphere. There is such a statement:

The bodies of multicellular plants and animals are cylindrical in shape.

This general statement identifies a particular feature of plants and animals that enhances their abilities to compete for space, food, and mates in the physical environment of Earth's surface.

A cylindrical body can be defined as a body having an approximately round or elliptical cross section and an easily identifiable longitudinal axis (Figure 1.1). In most cases to be mentioned the long axis will be greater than twice the diameter. There are, of course, exceptions—for example, lettuce plants, box tortoises, sea urchins, and sting rays—but for every exception there are thousands of species of cylindrical organisms. Some organisms—a rose bush and a spider monkey, for example—do not have an overall cylindrical shape, but the body itself is made up of cylindrical parts such as stems, roots, branches, torso, tail, and appendages. Other

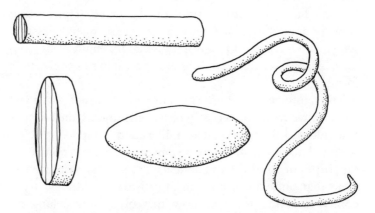

Figure 1.1 A cylindrical body has a circular or oval cross section and an identifiable longitudinal axis. Drawings by L. Croner.

organisms—for example, a sting ray and a butterfly—are noncylindrical as adults, but they grow and mature through cylindrical shapes during their development.

The statement that multicellular organisms are cylindrical in form leads to several questions whose answers might provide insight into the nature of all multicellular organisms:

1. What functional attributes does cylindrical shape confer on plant and animal bodies? What can cylindrical bodies do better than bodies of other shapes? These questions lead into a physical study of shapes and the mechanical properties of bodies having those shapes. I will explain how bodies can work on Earth, given the physical laws that allow us to predict that apples fall downward from trees while leaves go sideways when blown by the wind.

2. Do attributes of cylindrical shape confer selective advantage on their species? This question projects the physical plan of organisms into the evolutionary realm. The concept of evolution is the greatest general statement in biology, and it is therefore the most important biological context in which to set an explanation. The question bespeaks competition among organisms and the selective merits of different designs. It puts the single body into a context of interacting plants and animals as they toil in the physical world.

3. How did the cylindrical shape arise in the evolution of each species and in the development of individuals? This question really asks two historical questions in two very different time scales and opens the door for mechanical thinking in the areas of paleontology and developmental biology.

I will show how the posture and movements of organisms are physically based on their cylindrical body shape and how, in turn, cylindrical shape is unavoidably and even causally related to the mechanical properties and the arrangement of the structural materials within their bodies. Finally, a story will be told about how the cylindrical body shape may have evolved.

FORM, STRUCTURE, TIME, AND FUNCTION

This book is a study of the functional morphology of plants and animals. *Functional morphology* is the study of form, structure, and function and their relationships to one another. *Morphology* is the study of the form and structure of all things in the universe and in our imaginations. The *form* of a body is its external shape. The body's *structure* consists of the body's component parts and their shapes, spatial relationships, and connections. Although we may choose to study morphology alone, without reference to function, the study of biological form and structure does not, by itself, lead to a rich understanding of organisms.

Function refers to the dynamic physiological processes of organisms. It is dependent on form and structure over time. I will follow Picken (1960), who said that function is change in structure with time. He used the word *organization* to mean the combination of the structure and function of an organism, or the continuous set of processes of all its changes in structure over time.

Functions take place within or with respect to the bodies of organisms: to study function alone also leaves unacceptable gaps in our understanding of plants and animals. Because structure and process are not separate in organisms, they should not be separated in our thoughts and explanations. But not everything can be studied at once. Spoken language seems best suited for linear thought, and some temporary separation of topics makes many investigations easier than they would otherwise be. For example, it is difficult to view structure with a microscope at the same time that the oxygen concentration is being measured in a respiration experiment. Nevertheless, even though structure and function may be treated one at a time for analytical clarity, an understanding of the whole organism can arise only when structure and function are combined. Units of morphology (meter) and of function (meter/second) demonstrate that connection.

Biologists like to say that there are various constraints on organismic function, development, and evolution. Some say that the structure of an organism is a system of constraints. Thus, giraffes are constrained by their size and the length and shape of their legs from hopping like kangaroos, climbing trees like squirrels, or flying like birds. Even though this is a useful concept, it describes conditions that do not exist, and that is an endless task. What we need first is the direct approach: the complementary notion that structure permits the functions that exist. Giraffes can reach the leaves high in a tree without hopping, climbing, or flying, and it is the cylindrical form of their neck and legs that permits them to do so and still be light and supple enough to run from danger. In the process of setting forth the principles of mechanical design in organisms, it is at once helpful to name a structure and to say what that structure allows the organism (or its part) to do that could not be done without it. Later, in Chapter 5, I will use the concept of structurally permissible functions to account for the evolution of cylindrical body shape in organisms. I will show what new functional capabilities an organism acquires with each structural mutation and how functions permitted by macromolecular structures in turn permit and even cause the appearance of cylindrical body form.

The form of organisms is important for a number of reasons. First, it allows us to recognize the different kinds of plants and animals by sight and feel—two of our most important ways to learn about the world. We recognize a lobster and a fern by their shapes. We can see that shrimps and lobsters, while different from each other, are more like each other than either is like a fern. Thus, we classify, and the explanatory process is well on its way. Second, the form of an organism is essentially the shape of its mechanical support system: we readily recognize worms, beetles, cacti, fishes, and horses from their skeletons. This observation connects structure with mechanical support, and it sets the functional tone and direction of this book.

An organism's *mechanical support system* functions by accommodating forces: it may resist force, as our cranium does a sharp blow; it may transmit force, as our long bones do in their performance as levers; and it can absorb the energy of impinging force by stretching or bending, and thus it can avoid breakage, as do a sea gull's pinions in the wind, our skin in a pinch, and seaweeds in the surf. We can call the mechanical support system the *skeleton* as long as we include not only the rigid bones, shells, and other hard parts but also the flexible ligaments that hold them together and the pliant connective tissue sheaths that are stretched around the hydrostatic bodies of polyps, worms, and soft-bodied mollusks and sea cucumbers. In most animals, the skeleton is easily seen as a set of its own structural elements, such as bones and shells, separate from other systems. In the muscular hydrostats of mollusks, the turgid pith of wilting plants, and the rigid tissues of woody plants, no such clear separation occurs; it is awkward to speak of the skeleton of a squid, a buttercup, or an oak. Nevertheless, every organism has a mechanical support system. In plants it supports photosynthetic tissues and reproductive organs in positions where sunlight and gametes can get to them. In animals it is the basis of posture, and it reacts with muscular contraction to produce motion. The mechanical support system protects plants and animals from impact, breaking, and tearing.

PROPERTIES

Properties are measurable, conceptual attributes of structures. I will use properties to express the relationship between the cylindrical body form and its postural and locomotor functions.

> Reality is the embodiment of structure;
> Structures are the embodiment of properties;
> Properties are the embodiment of harmony;
> Harmony is the embodiment of congruity.

This interpretation (Haloun, 1951) of a verse by the fourth-century B.C. philosopher Kuan Tsi (quoted by Picken, 1960) is a good introduction to this line of thought. The connection of structure to harmonious function via properties is not a new idea, but it merits further contemplation.

The way to breathe life into the description of any object is to apply adjectives to it. A piece of cloth has little interest for us until we know whether it is starched, handwoven, salmon pink, translucent, knotted, torn, bespangled, or sodden. Properties are adjectival and adverbial descriptors that link structures and materials to specific functions. For example, the value for the stiffness of a material indicates the material's ability to support a constant or static load, whereas the value for the toughness indicates the material's ability to resist moving or repeated dynamic loads. Consideration of different properties focuses our attention sequentially on the various functions of a structure. For example, the optical properties of the cornea, the electrical properties of nerves, and the mechanical properties of skeletal structures help us to understand their primary functions. However, the mechanical properties of the cornea tell us about the role of pressure inside the eye in maintaining the precise shape that is necessary for sharp vision, and the electrical properties of bone are now being studied for their suspected role in stimulating the cells that are responsible for bone growth and repair. Properties are specific, often causal, relationships between structure and function: they add substance to our explanations.

Man-made buildings are large, dry, rectangular, rigid, and static. In comparison, plants and animals are small, damp, cylindrical, flexible, and dynamic. The cylindrical shape is rare in nonorganismic, natural objects: some crystals, stalactites, stalagmites, icicles, and lava tubes are cylindrical, but that is the complete list. In other words, nonliving natural structures and man-made structures are rarely cylindrical. This must mean that cylindrical body shape is a distinguishing property of multicellular life.

Mechanical properties are concepts created by engineers to describe those characteristics of materials and structural elements that make them useful in construction and fabrication. Mechanical properties are expressed in mathematical formulas and can be illustrated by graphs. A single example, tensile strength, will serve to introduce the subject here; others will be presented in Chapter 3. Figure 1.2 shows the results of an experiment wherein a tensile or stretching force was applied to a tendon from a dog's foot. The graph shows that the increase in length of the tendon was almost proportional to the increase in force right up to the end of the graph, at which point the tendon broke. The strength (S_{br}) of the tendon is the minimum force it takes to break it (F_{br}) divided by the cross-sectional area (A) of the tendon:

$$S_{br} = \frac{F_{br}}{A}.$$

This simple formula shows the relationships of the physical forces in our world to visible, tangible materials. Not only are mechanical properties links between physical laws and material objects, but they also express specific, quantitative

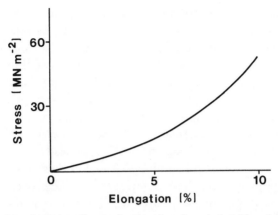

Figure 1.2 Stress–strain graph, showing the relationship of the elongation of a tendon from the foot of a dog to an applied tensile stress (force/cross-sectional area). MN, meganewtons; m, meters.

relationships between structures and functions. In the example given here, the tensile strength is the ratio of the force to the size of the object. This example also shows that properties are specific relationships among specified measurable features. Numerical magnitudes of properties tell us what functions a structure may have: strong materials may resist muscular action, transparent ones may transmit light to receptors, and so on.

Properties, like any other tool, can be misused. As biologists using engineering concepts, we must be sure we understand the assumptions and limits of each concept, and we must also be sure that the concept we are using is appropriate for our intended study. Because properties are most often measured on isolated tissues in a laboratory, we should never apply conclusions drawn from these results to the analysis of real, live organisms without first acknowledging that errors may have been made in making assumptions about conditions in and around the living organism. If knowledge of such properties is to contribute to our understanding of nature, properties must be seen in the context of the natural history of the species being studied. Laboratory studies on isolated morsels are informative, but the complex conditions surrounding their normal and abnormal function in life must also be learned if the properties are to relate structure and function to larger issues of ecology, behavior, and evolution. The process of gathering this information is complex and difficult, but the achievement of this goal leads to an understanding of plants and animals that makes them much more interesting and alive. The identification of the properties that are most important to a particular function requires all the intuitive insights and creative abilities of objective scientists. It is a vital step to be made in the earliest stages of planning a research project.

The mechanical properties described in engineering texts generally concern designs and design problems of man-made constructions such as towers, bridges, buildings, machines, and airplanes. Strength and stiffness are, for example, overwhelmingly important mechanical properties in the architecture of a rigid building that must not collapse or even sway visibly in the wind. Organisms, however, are small and flexible compared with buildings, and they do not contain materials as rigid as steel and glass. We are all designed to give before we break. Therefore, compliance, the inverse of stiffness, is the property to be emphasized in the materials of living organisms, and the breaking strain or deformation is often great and therefore more interesting than the breaking strength of many biological materials.

The most important role of properties in promoting our understanding of nature appears when the properties of all structural parts of a complex unit enable harmonious function. *Harmony* is the agreement of the placement and properties of the parts that promote function of a complex system.

At this time the units of harmony have not been identified, and it cannot be measured. But anyone who has ever sustained an injury to a joint in a finger, toe, arm, leg, back, neck, or jaw will know that regenerated tissues do not initially (and sometimes never) function as well as the original ones did before the injury. Harmony has been disturbed because the properties of the different materials in the joint are no longer well matched. Engineering designers work very hard indeed to assure compatibility of materials in the things they design. A hinge is an elegant solution to the simple problem of putting a flexible joint into the arm of a robot. But the problem becomes more complicated and harmonious function more difficult to achieve when electrical wires to motors and sensors, a water pipe, a gas line, and a cable to a motor that moves the hand must also bend around the joint, and when economy of space and material matter.

SCALES OF SIZE AND TIME

The size of a body significantly affects its function. When a mouse scratches its ear, its foot moves too fast for us to see. But when an elephant runs, we can easily watch the swing of its great legs. We can expect quite different degrees and rates of response to forces by structures that vary in size from that of mice to that of elephants. A functional difference between the mouse and the elephant is expressed by the difference in the respective rates of shortening of muscle cells in the leg.

The size range of living organisms and their molecular substructures is from 10^{-6} to 10^2 m, or 8 orders of magnitude in the total range (30 orders of magnitude) of all known structures from subatomic particles to galaxies. Bacteria are in the 10^{-6}-m range, the largest trees are 100 m tall (Figure 1.3).

If the first thing to notice about living organisms and their structures is size, the next thing to observe is how fast they

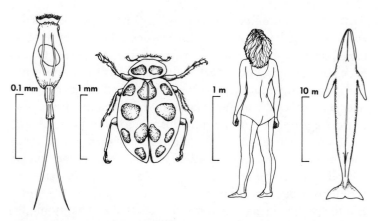

Figure 1.3 Organisms of different sizes reduced to the same size. Drawings by L. Croner and L. Srba.

move or change structure—and remember that changing structure is function. Physical laws describe the rates of motion of bodies in different size ranges. For example, gravity and viscosity of air or water differentially affect bodies of different sizes: the speed of motion is limited disproportionately according to the size of the body. In water, for spherical bodies less than 1 mm in diameter, drag (the force that opposes motion) is proportional to velocity, whereas for bodies greater than 1 mm in diameter, drag is no longer simply proportional to velocity. So, at some point, further increase in size of aquatic organisms will cause a disproportionately greater drag resistance to their motion. Additionally, tiny animals that swim or fly get little benefit from being streamlined. These differences in response to physical factors by bodies of different sizes (molecules, cells, tissues, organs, and organisms) contribute to our understanding of properties of organisms and their parts.

The rates of movement of organisms and their parts are greatly affected by the fact that they are wet. For macromolecules in sopping wet extracellular materials such as the collagenous connective tissues in animal bodies, the drag on intra- and extracellular protein and polysaccharide molecules has a damping effect on deforming forces and on the rate and degree of elastic recoil. Damp structures resist rapid deformation and are slower to recoil than dry ones.

Molecular motion is too fast to be seen with the unaided eye. But changes in shape of the largest macromolecular complexes, such as the sliding of actin and myosin molecules along one another in muscle contraction, are visible. With the help of a microscope, we can see macromolecular structures, such as flagella, making waves. Large, whole cells are even more easily observed in motion: although flight muscle cells of the fruit fly *Drosophila* contract and are reextended in less than a hundredth of a second, muscle cells in a whale take up to 10 s to wag the tail once.

Because change of structure is function, the rate at which structure changes is of great functional importance. Organisms function at three basic speeds. The fastest functions are the *physiological changes,* many of which are chemical and therefore extremely fast, for example, the light reaction in photosynthesis or the transmission of an impulse from a nerve cell terminal across the synapse to the membrane of a muscle cell. In this category of fast functions I will include *motions,* although they are not nearly so fast as chemical reactions; nevertheless they are changes in form and they are functional (Figure 1.4A). The next slower structural changes are those that occur over much of the life span of an organism, namely, *developmental changes,* which include such dramatic reorganizations as metamorphosis (Figure 1.4B). And finally, the slowest of all are the *evolutionary changes* that take

A

B

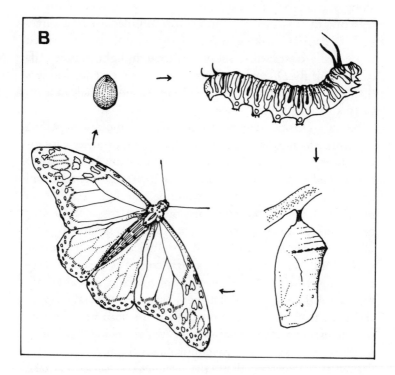

C

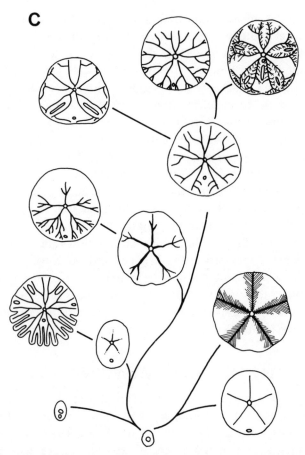

Figure 1.4 Changes of form in three time scales. (A) Physiological changes can happen quickly—in seconds or fractions of seconds. (B) Developmental changes happen more slowly and can take minutes or years. (C) Evolutionary changes are the slowest: they take generations. A and B by J. Harmon; C by L. Srba.

generations to accomplish (Figure 1.4C). But, remember, each level in the time scale pertains to structures at all levels in the size scale from molecular through the whole organism.

Cylindrical bodies obey the rules of scale. In fact, consideration of the size scale was the original stimulus for the appreciation of the mechanical implications of the cylindrical form in large organisms. Borelli (1680) understood that bodies increase their mass with the cube of the radius, whereas the strength of a supporting cylindrical stem or leg increases with the square of its radius. Thus, if terrestrial plants and animals are to achieve great size, they cannot just get bigger without changing shape—the radius of cylindrical supporting parts must increase faster than the length.

Also, whereas the cylindrical shapes of fishes and seaweeds contribute to their streamlined shapes, tiny bodies of protists, moving at much slower speeds, march to different physical drummers that do not require the small bodies to be cylindrical. Consequently, protists have been specifically excluded from my account—they should be the focus of another book.

THE MECHANICAL VIEWPOINT

One advantage of making a mechanical study of organismic form is the relative simplicity with which it can be done. With only a meter stick or its variant, a balance or spring scale, and a stopwatch or calendar, we can measure structures and forces and their rates of change. Then we can calculate the values of properties that will enable us to create a mechanical explanation of how a particular plant or animal stands or moves. It is this line of investigation that shows the cylindrical shape of organisms to be the most suitable one.

Stance and motion enable organisms to compete for environmental resources. Posture affects competition for sunlight and flow-delivered food and other resources. Movement affects feeding, reproduction, and behavior. Successful organisms put the resource catcher—the chloroplast or the mouth—where it will most often contact the resource. If, like a tree or a coral, the organism is nonmotile, its branches must extend out further than its neighbors to ensure its harvest. Reaching out presents a problem in mechanical design, and I will show that the most efficient use of materials in support systems that reach out occurs in cylindrical bodies. If, like a shrimp, a newt, or a bird, the organism can move through the environment, it will be more efficient (it will expend less energy) in meeting new food items if it is cylindrical, because cylindrical bodies are basically streamlined. Finally, for animals to move about on land and in the air, appendages supported by stiff cylindrical rods are the best designs yet for lever-like locomotor organs.

2
THE MECHANICS
OF SHAPE

～⁓✦⁓～

The whole organism is an important unit of structure and, as such, has its own unique functions. The context for the present study of an organism's function is the physical environment in which organisms function. The major mechanical factors are gravity and the viscosity of wind and flowing water. It is within this context that natural selection takes place.

Posture and movement are the organismic functions that deal with gravity and fluid flow. Posture enables leaves to intercept sunlight, flowers to be pollinated by wind, insects, or bats, and roots to reach water. It enables animals to perceive food, mates, and predators. The movements of animals enable them to change posture, to scratch an ear, to escape or to capture, and to carry out their activities at different places. Movement is behavior, whether it is feeding by one, breeding by two, or migrating by a million. Although the movements of plants are more subtle, they enable some flowers to follow the sun, some plants to catch flies, and tumbleweeds and milkweeds to deliver seeds downwind. Posture and movement are mechanical functions permitted, limited, and controlled by body form and by the substructure and materials of the body.

In the next three chapters I will investigate mechanical design features of cylindrical bodies and their substructure. I will show how the cylindrical shape of structural elements (back, thigh, and jawbones, palm fronds, lobster and mollusk shells, ligaments, exoskeletons, and fluid-filled cavities) contributes to their mechanical properties and to the behavioral capabilities of the bodies they constitute. Having properly introduced the cylindrical skeletal element, I will then show what functions cylindrical bodies do better than bodies of other shapes do and which of these functions might be advantageous to plants and animals. In Chapter 3, I will discuss the kinds of materials of which cylindrical structural elements are made. The bone of bones, the fat and sinew of blubber, and the fibrous, pithy, and woody tissues of plants all have substructures that contribute to their usefulness in cylindrical supportive elements. Finally, in Chapter 4, I will consider whole organisms as mechanical systems of shaped elements and will show how the arrangements of cylindrical elements enable cylindrical organisms to cope with each other and with the mechanical rigors of the environment.

ONE-, TWO-, AND THREE-DIMENSIONAL SHAPES

Just as organisms have shapes that indicate their functions, so do the components of their support systems. Stems, nut husks, bones, shells, and tendons are as instantly recognizable by their shapes as are bricks, I beams, nails, and hinges. Man-made items have been improved over a relatively short time to serve their mechanical functions as well as they can, whereas natural selection has taken very much longer to improve the design of mechanical structures in nature. Why, then, are leg bones cylindrical, shoulder blades platelike, and skulls domed? Why are the stems and roots of so many plants cylindrical? It seems sensible to assume that the shape of each skeletal element is appropriate for its mechanical function, because, as Buckminster Fuller continually reminded us, the physical rules that prescribe the best shape for each mechanical function are the same for Nature as for human designers.

Linear elements are virtually *one-dimensional;* for example, fibers, strings, ropes, and cords are linear elements. They are really just very slender cylinders; but they can be distinguished from the three-dimensional cylinders, to be treated later, on the basis of their substructure and a functional property that arises from the substructure. Linear elements are parallel aggregates of long, thin cells, like those in muscles and stems of hanging fruits, or of macromolecules, like those in cotton fibers, tendons, and ligaments; and they all function in tension—by being pulled on (Figure 2.1). In fact, anywhere one finds parallel aggregates of long, thin fibers in natural or in man-made materials and structures, one can safely assume that the aggregate functions by accommodating tension. That assumption predicts the solution to the riddle, "What does that structure *do?*" Parallel, fibrous aggregates are one of the most common patterns of recognizable order to be found in plants and animals and are present in

Figure 2.1 Pliant structural elements that function in tension. Linear (one-dimensional) elements include (A) stems of fruits (cannonball tree), (B) tendons and some muscles, and (C) silk threads in spiders' webs. Two-dimensional elements include (C) the entire web, (D) the stretched skin of a puffed-up puffer fish, (E) *Physalia,* the Portuguese man o'war, and (F) the cross section of a shark. In F, the eccentric profiles are collagenous myosepta. Each myoseptum is a cone-shaped membrane. Note that where they meet the skin, they fan out and distribute their pull on the skin. Conversely, septa converge to form horizontal septa that meet the rigid backbone at focal points. Drawings by L. Srba (A,B,C) and G. Minnich (D,E,F).

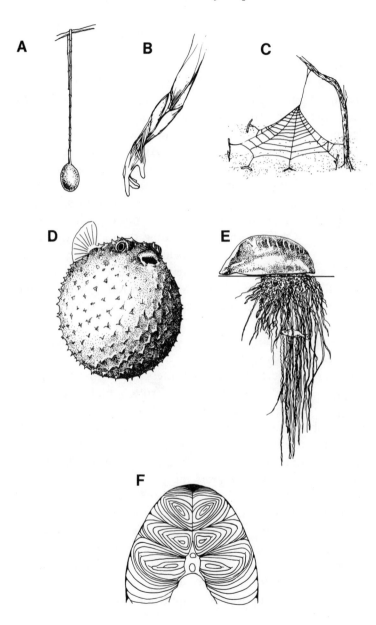

virtually every species. And such a package is almost inevitably a structure that is pulled on.

Notice that I have stated that these linear elements *accommodate* tension, not just *resist* it. When a string and a rubber band are pulled on, they are certainly resisting tension, but they do so neither by the same molecular mechanism nor to the same degree (see *stiffness* in Chapter 3) and thus will not be useful for the same mechanical functions. Apple stems and the tendons of human hands and arms function by being stiff when pulled on. They do not stretch very much. The apple stays tightly on the branch and the tendons resist tension and transmit the forces and the displacement of muscle contraction to the bones. On the other hand, some tendons in the flight systems of insects function by being stretched; then they recoil elastically to accelerate the flapping wing. So, to indicate that all these items are subjected to tension, but that some stretch more than others, I have said that they accommodate tensile forces.

Among the most elegant of tensile cords are the silken threads spun by spiders and caterpillars and some of their relatives (Figure 2.1). Tensile stiffness of the thread supports a spider as she lets herself down from a high perch. The tension a spider puts into a dragline on the ground or into a web enables the spider to detect vibrations made by the animals that stumble into these structures, because vibrations are simply minute and frequent changes in the tension of the thread. The spider interprets the patterns of vibrations and decides whether to attack, to approach amorously, or to remain hidden. And, of course, both the strength and great extensibility of the threads in the web are responsible for delaying the escape of prey until the spider can wrap it in more tensile threads. Spiders are masters of the tensile thread.

Tension is useful. And it is simple, especially when compared with its opposite—compression (squashing)—or its devious cousins—bending and torsion (twisting). A slender rod made of one material can support the same maximum load in tension, no matter how long the rod is (Figure 2.2). But the maximum load the same rod can support in compression, bending, or torsion, decreases as the rod becomes longer. From this observation comes the notion that pound for pound, supportive elements put only in tension can save the builder money for materials and, if the construction is a mobile one, for transport. Tensile elements can be slender and graceful: suspension bridges are airy, soaring structures that appear not to have enough material in them to carry heavy traffic when compared with bridges built of stones set in semicircular arches. Buckminster Fuller was a champion of tension and never tired of telling us that, in the design of buildings, the more loads one can support in tension, the lighter and cheaper, more soaring, and more graceful the building can be.

Two-dimensional elements, or *sheets,* are either soft, flexible membranes, such as the leaves on a tobacco plant and the skin in the wing of a bat, or rigid plates, such as the back or breastplate of a turtle and the jawbone of an ass (Figure 2.3). Sheets accommodate tension in any direction in the plane of the sheet (Figures 2.1D,E and 2.4). When you blow up a balloon or eat too much dinner, the balloon or your stomach is stretched in all directions, while the air or dinner is compressed inside. Because the materials of which most sheets are built have constant volume, the sheets must become thinner as they are stretched (Figure 2.4). Another way to make something thinner is to squash or compress it. In other words, tension in the vessel wall imposes compression on its contents. There is compression normal to the sheet and tension in the sheet.

Thus, a sheet can be stretched either by pulling on its edges or by applying pressure from one or both sides. *Pressure* is the

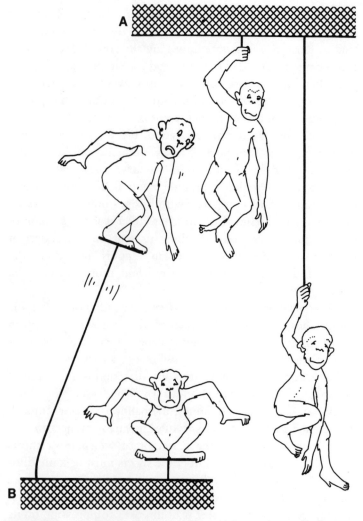

Figure 2.2 The ability of a piece of material to resist tensile forces (A) is independent of its length, but the ability of the piece to resist compressive forces (B) is inversely related to its length, because of its tendency to buckle. Drawing by N. Meith.

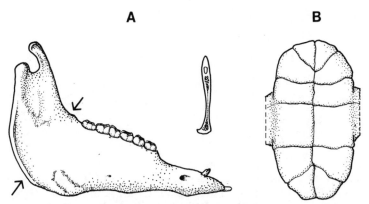

Figure 2.3 Rigid, two-dimensional sheets. (A) Lateral view and cross section (*at arrows*) of the jawbone of an ass. (B) Breastplate, or plastron, of a box turtle. Drawings by L. Croner.

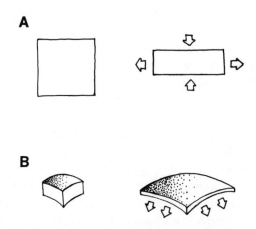

Figure 2.4 (A) When the volume of a material remains constant, it contracts in the direction normal to the stretch. (B) Stretching a sheet in two directions in the plane of the sheet requires a decrease in its thickness to maintain its constant volume. Drawings by J. Harmon.

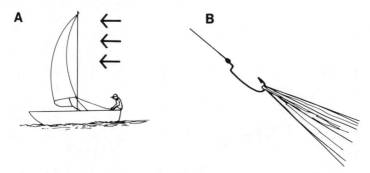

Figure 2.5 Distributed loads and point loads. (A) Soft membranes are well suited to accept a distributed load, such as air pressure. (B) Membranes are not well suited to accept point loads, which may kink, tear, or pierce the membranes. Drawings by L. Srba.

force per unit area acting perpendicular to a sheet. A flexible membrane bulges when the pressure acting on it is greatest on its concave side. But a stiff sheet, such as a human cranium or a snail shell, can resist compression from either outside or inside.

Pressure, when transmitted by a fluid (either a gas or a liquid), applies a distributed load to any surrounding sheet; the force is applied equally over the entire area (Figure 2.5A). Flexible membranes are well suited to accommodate distributed loads, but they are not well suited to take point loads—one does not expect to see a linear tendon attached to a point in the middle of a soft membrane. A pull on the tendon would produce an awkward, tentlike deformation that dangerously concentrates the force and threatens to tear the membrane (Figure 2.5B). Because membranes are best suited to take distributed loads, when one membrane pulls on another, the tensile attachment should be fanned out, to distribute the load over a wider area of the receiving membrane. In the cross section of a fish shown in Figure 2.1F, note the broad, fanned-out attachments of membranous myosepta to the membranous skin and the narrow attachments of the myosepta to the rigid backbone.

Rigid plates are the stock in trade for many familiar man-made structures, such as cars, computers, and mixing bowls. And there are good reasons for building houses and other buildings out of flat plates: it is more convenient to walk on a flat floor than on a pleated or curved one, and flat plates are easier and cheaper to build. Nevertheless, flat plates are intrinsically weak. Unless they are made very stiff by folding or by adding additional material, they tend to become nonflat when they are loaded: floors and flat, horizontal roofs sag with people and snow loads. Notice that modern cars and aircraft are becoming increasingly less flat-sided; and flat-walled pipes are never used to carry water, gas, and oil under pressure.

Flat plates in nature are rare and usually represent a compromise between the energy and materials saved in making them and some real advantage. For example, jawbones of horses and other large mammals that grind grain and fibrous vegetable matter for food are flat (Figure 2.3A). Their great lateral areas provide room for the attachment of enormous chewing muscles, and the action of those chewing muscles loads the bone only in the plane of the plate, a direction in which it is very strong indeed. But jawbones are much weaker in the direction normal to the plate.

A flat plate can be stiffened by being pleated or curved into half-cylindrical, saddle, or bowl shapes. Human beings have learned to fold roofs, imitating the form of some insect wings and palm leaves (Figure 2.6). And shells, skulls, and the crisp exoskeletons of crabs and cockroaches are examples of cylindrical and bowl-shaped plates. Most insect wings and leaves are broad and flat, but stiff, because they are reinforced by veins or ribs.

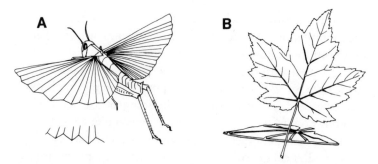

Figure 2.6 Thin sheets can be stiffened by folds (A), as in a locust's wing, or by the addition of ribs (B), as in a maple leaf. Drawings by G. Minnich.

Three-dimensional elements are lumps. If they are made of soft material, they transmit pressures, but they may give when subjected to tensile forces. If they are made of rigid materials, they can transmit any kind of force in any direction. When we stand or sit upright, the vertebrae in our backbones are compressed across their flat ends by the weight or our torso; and they are bent and twisted by our postural antics and pulled in many directions by muscles and ligaments.

When lumps are longer in one dimension than in others, they are cylinders. Leg and arm bones and ribs and most roots and stems are three-dimensional structural elements called *beams* (rather than cylinders) by engineers. The important thing to realize about beams in man-made and natural structures is that the forces they meet in the natural environment tend to bend and twist them. Compared with tension and compression, bending and twisting (or torsion) are complex and thus potentially dangerous deformations. The variation that occurs in the structure of cylindrical stems, branches, bodies, and appendages and that reflects this complexity reveals a great deal about the postural rigidity and locomotor capabilities of the organisms themselves. The variety of these forms in nature is great; but the fact that most bodies and appendages are still recognizably cylindrical is testimony to the usefulness of the cylindrical form.

CROSS-SECTIONAL SHAPE AND MECHANICAL PROPERTIES

Beam theory deals with the mechanical significance of the length and cross-sectional shapes of beams (Figure 2.7). Beams are usually thought of as rigid, long, thin structural elements with rectangular, T-, or I-shaped cross sections. But a beam really may have any cross-sectional shape. Beam theory enables engineers to design the sectional shapes of beams for specific purposes. Given the loads to be sustained and the material to be used, the theory enables us to calculate the least amount of material that will do the job, or to determine the appropriate length and diameter. Given the length required by the whole design, it enables us to determine the minimum diameter of the given material that will support the load, and the shapes of the beam's cross section that are appropriate or even allowed. If it is a hollow beam, we can determine how thick the wall must be with respect to the diameter of the beam. And, finally, if more than one material can be used, the materials' mechanical properties are considered in the determination of how they should be distributed across the cross section of the beam to achieve compatability of materials when the beam is bent.

Clearly, beam theory is an instructive and useful theory for both analytical scientists and design engineers. But, like all scientific theories, it is based on assumptions; and, like the assumptions behind almost all physical theories, they are rarely totally justified when applied to biological systems.

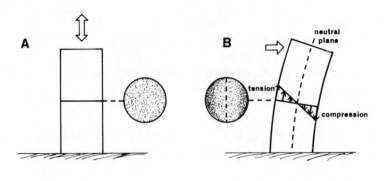

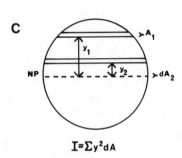

$$I = \Sigma y^2 dA$$

Figure 2.7 Beam theory. (A) Simple tensile or compressive forces (*hollow arrows and even stipple*) are evenly distributed over the cross section of a beam. (B) Tensile stresses (*solid arrows and uneven stipple*) are greatest on the convex side, and compressive stresses (*solid arrows*) are greatest on the concave side in a bent beam. The hollow arrow represents the bending force. Tensile and compressive stresses decrease to zero at the neutral plane (*dashed line*) in the middle of the bent beam. (C) The second moment of area, *I*, of a beam's cross section says that the contribution of an increment of area (*dA*) to the stiffness of the beam varies with the square of *y*, its distance from the neutral plane (*dashed line*). So, a hollow beam is stiffer than a solid one of the same sectional area. Drawings by L. Srba.

For example, beam theory is based on the assumption that the material of the beam is rigid and homogeneous and that flat transverse planes in the nonloaded beam will still be flat in the loaded or bent beam. Although some of the cylindrical beams in organisms are rigid, none is homogeneous; and no one has rigorously determined that flat planes stay flat when biological beams are bent.

In a beam subjected to simple tension or compression along its axis, the stresses are evenly distributed across the cross-sectional area of the beam (Figure 2.7A). A beam in tension is like a linear fiber, a tendon, or a guy wire: as far as its ability to accommodate tension is concerned, there is virtually no limit to how long the fiber can be (Figure 2.2). Engineers designing large structures must worry about the weight of very long cables, but only a few natural tensile structures, for example, the great lianas (vines) in tropical forests, are faced with the problem of self-weight brought about by extreme lengths. On the other hand, a beam can only accommodate axial compression as long as the compressive stress is precisely parallel to the beam's long axis. Any movement of the compressive force away from the axis causes the beam to bend (Figure 2.7B). If you grab a meter stick by both ends and compress it by bringing your hands closer together, the stick seems stiff until your awkwardness shifts the force axis a little to one side of the stick. Then, not only does the stick bend, but it does so suddenly, and there is a real possibility of its breaking. This example illustrates the practical importance of beam length. The directions of forces in the habitat of an organism are always changing, so long, thin compressive beams will undoubtedly be bent. In other words, beams resist compression in inverse proportion to their length. Short, fat beams can resist compression. All the others get bent.

Bending leads to a complex distribution of forces in a beam, because it puts the convex side of the beam in tension and the concave side in compression. As Figure 2.7B shows, the stresses are greatest at the convex and concave surfaces and decrease to zero in the middle of the beam. All the points in the middle of a beam lie in the so-called *neutral plane,* where stress is zero. In other words, stresses are distributed unevenly across the cross section of a bent beam. The material in the center half of the beam is not taking its fair share of the bending load, and stresses are concentrated in the peripheral material. A beam that can easily withstand a given load under simple compression or tension may fail on its convex, tensile side when it is bent, because of the concentration of stresses.

The *flexural stiffness* of a beam is given by the engineering expression *EI*. The quantity *E* is the *stiffness* of the material: it represents the amount of force per unit cross-sectional area required to stretch the material a given amount. Stiffness is a property that arises from chemical bond energies and orientations characteristic of the material. The quantity *I* is a morphological index, called the *second moment of area,* that represents the contribution of each bit of material, in each position in the cross section of the beam, to the beam's resistance to bending. Its formula is

$$I = \Sigma \, y^2 dA,$$

where *y* is the distance from the neutral plane of each element of area (Figure 2.7C). The relationship between the radius and the second moment of area is dramatic. For the material on the outside of a cylinder (or all the material in the wall of a tube), *y* approaches the value of *r*. Because the area of a circle is πr^2, and the formula for *I* multiplies y^2 by r^2, *I* will vary with the fourth power of the radius. In other words, increasing the diameter of a cylindrical beam a tiny bit increases its flexural stiffness a lot. Fibers, wires, and other very thin beams are very flexible simply because their radius is small, no matter how stiff the material is. Obviously, the radius of a beam is enormously important to its flexural stiffness.

Figure 2.7B and C show how important the outermost material is in resisting bending and how relatively unimportant the inner material is. Really, the only function of the inner material is to maintain a large radius by keeping the outer material away from the neutral plane, especially when the beam is bent. This reasoning lies behind the design of I beams, which are used in the construction of buildings and other large static structures in which the direction of the forces is predictable: one can count on gravity always to pull toward the center of the earth, so one orients the I beam with its maximum I parallel to the pull of gravity. In situations where forces may come from any direction, the mechanically important outer material can be arranged continuously—in a tubular form.

Tubes are hollow cylinders. Not only is the hollow, circular shape the best one for cylinders that must resist omnidirectional forces, a circular cylinder of uniform wall thickness also is most economical with respect to materials. Stems of bamboo and other grasses, lilies, and dandelions and the long bones of fore and hind limbs, feather shafts, and beetles' leg segments are stiff, hollow cylinders that resist bending. Snakes, worms, sea anemones, guts, arteries, honeysuckle vines, and some seaweeds are cylindrical, but they are flexible.

In the strategic matter of whether cylinders have high or low flexural stiffness, there are two properties to consider: E and I. Stiff cylinders may be stiff because they are made of the stiffest materials available (high E), or they may have the best sectional shape for their situation (high I). Bamboo, for example, uses both high E material and high I shape at the rigid base of the plant. Flexible cylinders may either be made of pliant materials or have reduced diameter or both. The sea anemone has a large diameter (large I) which can accommodate a large lunch; but its body-wall material is a soft, stretchy collagenous connective tissue called mesoglea, which has a low E. When anemones bend, they contract circular muscles over a short length of their columnar body, thus locally reducing the diameter and I, and making a flexible "joint" in the less flexible body. Legs of arthropods are hollow, jointed, cylinders of light, rigid material. Flexible joints in the legs occur where a pliant but nonstretchy material, giving a lower E, and a smaller diameter, giving a lower I, are combined. Indeed, throughout the plant and animal kingdoms, many such combinations of material stiffness and second moment of area occur.

Probably no biological cylinder is homogeneous across its cross section. In flexurally stiff structures, such as a corn stalk and our leg bones, the stiffer material is concentrated at the outer surface of the cylinder. The incredibly flexible stipes (a technical word for the stems of seaweeds) of giant kelps such as the bull kelp, *Nereocystis luetkeana,* grow to lengths exceeding 10 m, and much of the stipe has a diameter of less than 5 cm. The stiffest material in the stipe is as close to the center of the section as possible, where it contributes very little to flexural stiffness. In other words, it keeps flexural stiffness low.

I is an important factor in the mechanical properties of beams. From considerations of I we can conclude that a beam with a round section is equally stiff on all sides and that a beam with an elliptical section is stiffer in the direction of the major axis of the ellipse. Depending on how small the minor axis is, the beam may be very flexible in that direction. The common purple sea fans (*Gorgonia flabellum*) of the Caribbean are often a meter tall. The cross section at the base of such a fan is an exaggerated ellipse, and the gentlest surge causes the fan to bend at the base in the direction normal to the fan. However, the base is so stiff in the plane of the fan that a force applied in that plane will tear the fan off the reef before bending it in that direction. In this sense, feathers and leaves might be thought of as even more extreme cases of elliptical beams. Leaves, insect wings, and any other flat plate or ribbon can be made more rigid by being folded like a foldable hand fan (Figure 2.6). This arrangement increases their second moment of area normal to the fan.

Other simple cross-sectional shapes confer interesting properties to beams. Figure 2.8 shows values of I for shapes that have the same cross-sectional area. The I of a hollow beam is dramatically greater than that of a solid one. The petioles (leaf bases) of some plants are triangular in section, and stiff material is concentrated along the flat top of the

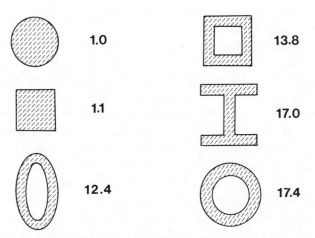

Figure 2.8 Values of second moment of area, I, for various cross-sectional shapes of equal area. These values are for beams being bent up or down, parallel to the sides of the page. Drawings by L. Croner.

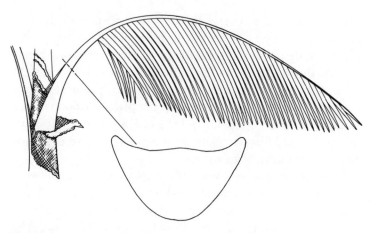

Figure 2.9 The base of the leaf of the coconut palm is almost triangular in shape. The leaf base is stiff in bending (high I), but it twists easily (low J). Drawing by G. Minnich.

petiole, where gravitational forces cause the leaf's mass to apply the greatest tension in bending.

Stiffness against twisting, or *torsional stiffness,* is given by GJ, where G is another kind of stiffness (the stiffness to shearing forces, see Chapter 3) and is analogous to E, and J is the polar moment of area, analogous to I, and indicates the importance to torsional stiffness of each element of the cross-sectional area. Round beams resist twisting forces according to the fourth power of their radius. Beams with elliptical sections, however, are more torsionally compliant. In Figure 2.9, the sectional shape of the base of the coconut frond (leaf) is an exaggerated isosceles triangle. A wind will not easily bend the frond base to either side (high I), but it will be more easily bent toward and away from the trunk (lower I); and it will be easily twisted (low J). The twisting of the petiole allows the great blade of long leaflets to orient normal to the wind. In this position, the petiole can bend with the wind a half meter or so from the trunk. Bending to either side at midfrond allows the blade to orient with the wind and thereby to decrease the wind drag tending to tear it off the tree. Allowing bending only away from the trunk prevents leaves from entangling and abrading each other.

The cylindrical body shape can be a simple beam with similar properties in all directions normal to it; an earthworm has such a shape. But with simple variation in diameter and, if hollow, in wall thickness, the cylinder can easily become a streamlined body of differential flexibility and twistability at various points along its length. Heads and bodies of birds, for example, are quite rigid, whereas their necks and tails are flexible. Certainly there are materials of different stiffness (E) found in these places, but subtle local differences in wall thickness or cylinder diameter (and therefore differences in I) have large consequences for the flexibility of the body.

3

MATERIALS AND THEIR MECHANICAL PROPERTIES

The most striking difference between man-made materials and the structural materials in organisms is that biomaterials are wet. Many biomaterials are living, for example, wood, tendon, skin, and bone; some are not, for example, coral, shell, scales, claws, hair, and the cuticles of plants, insects, and other arthropods. High water content in a material gives that material a wide range of properties that are not simple to predict. The water may actually flow in the material as it is stretched or compressed, and this flow will cause some properties, for example, the stiffness, to change with time after a force is applied.

Another vivid difference between man-made materials and biomaterials involves their degrees of complexity: biomaterials are much more complex than the synthetic ones. For example, fiberglass is a fairly complex synthetic material; it is a mixture of glass fibers embedded in a plastic matrix. Its mechanical properties depend on the amount of parallel orientation of the fibers, and its ability to be useful in flexible fishing rods depends on the low stiffness of the plastic. On the other hand, the complexity of bone is mind-boggling; it contains submicroscopic, plate-shaped crystals of a mineral (calcium hydroxyphosphate), collagen fibers, and coiled gly-coproteins and other mucuslike molecules. There is also a small but significant amount of fat in bone. The crystals and fibers may be oriented, and any of the components may be present in different concentrations in different microzones of the material. Properties of bone are always different in different directions. Its properties vary from species to species; and, within a species, each property may vary with the age and health of the individual as well as with the location of the bone in the body and the site on the bone. Properties vary with chemical composition and the orientation and spacing of the chemical components. Bone also has minute holes in it (where the cells live), and these holes also affect the properties. And there are at least three more hierarchical orders of organization in which bone structure and properties vary.

Structural materials and their mechanical properties are a major determinant of the postural and behavioral capabilities of plants and animals. But an understanding of the physical determinants of the mechanical properties of biomaterials requires a knowledge of fluid, polymer, and solid mechanics. A taste of this will be given in this chapter.

STRESS AND STRAIN

Mechanical properties of tissues depend on the substructure of their constituent materials. Shape, size, and orientation of molecules, microscopic fibers, crystals, and even dust particles affect material properties. In highly deformable materials, components are not tightly attached and slide past one another. Ultimately, whether a material holds or breaks under a load depends on the strength of its interatomic bonds.

No matter what material is being considered, every force applied to it will cause a deformation. This observation is one of the great physical truths about the nature of materials in this universe. If you pinch a steel girder, you may not be able to see the dent under your finger, but rest assured that the iron atoms under your fingers have indeed moved a little bit closer to one another. Because you probably only stretched or bent or compressed interatomic bonds and did not break any, the instant you released the girder, the iron atoms popped back into place elastically. *Elasticity* is a property of all solid materials, and it has two components: one is the ability of the material to be deformed by a force, and the other is the tendency for the deformed material to regain its original shape when force is removed. *Deformability* (stretchiness, squashiness, flexibility, and twistability) and *recoil,* whether partial (as in bread dough) or complete (as in springs, rubber bands, and bird feathers) are basic properties that we can measure and understand.

A standard way to learn something about the mechanical properties of a material is to subject a sample of it to a tensile test, that is, to pull on it with a known force and measure the amount of deformation (Figure 3.1A). To make the information revealed by this test comparable to results of similar tests

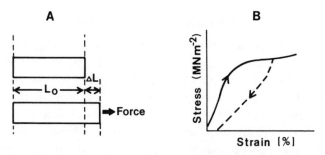

Figure 3.1   (A) In a force–extension test, a tensile force causes a piece of material of original length L_o to extend by ΔL. (B) The results of the test are shown in a stress–strain graph. Stress = force/cross-sectional area and strain = $\Delta L/L_o$. Drawings by L. Blalock.

on other pieces of materials, the force is divided by the cross-sectional area (F/A) of the sample to give the *stress,* in units of newtons per square meter (Nm^{-2}); and the amount of stretch (change in length, $L_o - \Delta L$) is divided by the original length (L_o) of the sample, giving the dimensionless *strain,* ($L_o - \Delta L)/L_o$, usually expressed as a percentage of the original length. A *stress–strain curve* (Figure 3.1B) shows the relationship between these quantities. There is morphological information in both quantities: stress is proportional to an area and strain is the ratio of two lengths. Thus, these quantities show a mathematical relationship between specific morphological and functional factors.

Figure 3.1B shows the results of a tensile test on a piece of bone stretched to the breaking point. In the test, a force is applied to the bone in the direction of its long axis. As the force increases, the stress increases, because, in rigid materials such as bone, the cross-sectional area does not change appreciably over the short extension it takes to break them. The first part of the graph shows that, as the stress increases, the strain increases. The steeply ascending part of the curve is called the *elastic region.* The slope of the elastic region (stress/strain) is the *stiffness* (called *elastic modulus* by engineers) of the material and has the same units as stress: Nm^{-2}.

If the force, and thus the stress, is removed during this part of the test, the curve showing the strains as the load is removed would be superimposed on the loading curve, and the unloaded bone would be the same length after the test as it was before. The sharp bend at the top of the curve is called the *yield point.* If the stress is removed after this part of the test, the unloading curve (dashed line) shows that permanent deformation has occurred. For this reason, the part of the curve beyond the yield point is called the *plastic region.*

Materials such as bone normally function in the steep, elastic region of this curve, but the deformation allowed by the horizontal, plastic region serves as a safety factor contributing to the toughness of bones. The end of the curve shows the conditions present when the material actually breaks. The value of the stress at the breaking point is referred to as the bone's *tensile strength,* or *breaking stress;* and the value of the strain at the breaking point is called the *breaking strain.* In general, very strong materials (high tensile strength) are not very stretchy, and stretchy ones (high breaking strain) are seldom strong.

DEFORMABILITY, TOUGHNESS, AND BRITTLENESS

Another important property can be gleaned from the stress–strain curve. The area under the stress–strain curve is a function of the product of the stress and the strain:

$$\text{stress} \times \text{strain} = \frac{F}{A} \times \frac{\Delta L}{L_o} = \frac{\text{work}}{\text{volume}}.$$

Stress is a force term and strain is a change in length, or distance. Force times distance is work or energy. So the area under the curve is a measure of the amount of work or energy per unit volume put into the material at any point on the stress–strain curve. At the breaking point, the area under the curve tells how much energy it took to break the object: this is called *toughness,* obviously a most useful property to consider. Weak, stretchy materials like skin may take as much energy per unit volume of the material to break as do strong, nonstretchy materials like bone. This fact illustrates an important thing about complex properties: there is more than one way to control them. A material may be tough by being either strong or stretchy or both.

Consider the stress on a leg tendon of a running cow and the stress on a giant kelp stalk whose floating fronds are

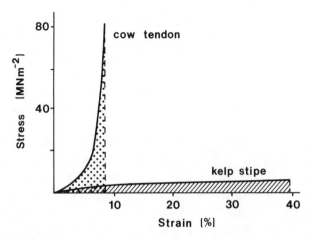

Figure 3.2 Stress–strain graphs for a leg tendon from a cow and the stipe (stem) of a giant kelp. Breaking stress is much greater for the tendon, but the kelp stretches much more than the tendon before it breaks. As a result, both tissues require about the same amount of energy to break them, as indicated by the areas under the curves. Drawing by L. Blalock.

pulled on by storm waves while the plant is anchored to the seafloor. Figure 3.2 shows that the material in the cow's tendon is very strong and that the kelp stalk is much weaker than tendon. But the breaking strain of the kelp is ten times that of bone, so the toughnesses of the two strikingly different tissues are similar. Notice also that bone is much stiffer (higher slope of the elastic region) than kelp stipe.

A running step of a cow puts high stress (the entire weight of the cow on the cross section of the tendons) for a tenth of a second. The tendon must be stiff if the locomotion is to be at all efficient. A large storm wave pulling on the fronds of the kelp develops a shearing drag force that increases from zero to a maximum in 4 or 5 s and then fades again. Time is the important thing here. It takes time to stretch a 20-m long kelp stipe to the 45 percent extension required to break it, and there just is not enough time in the passing of even a huge wave (10 s) to allow for this much deformation. Kelp cannot gallop and cows are not tethered to the seafloor during storms. Tendon is stiff and strong, but kelp is weak and stretchy. Both are exposed to drastic mechanical conditions, and both are tough.

A property that often accompanies high stiffness of highly mineralized materials like bone, shell, and sea urchin stereom is *brittleness;* the ability of a material to resist the propagation of cracks. Brittleness is the converse, but not the inverse, of toughness. A glass rod is brittle; it breaks because a crack has propagated across it at a very high velocity. Homogeneous materials are often brittle, and the regular array of chemical bonds holding them together facilitates the flow of crack-making energy from neighboring broken bonds to the tip of the advancing crack. Figure 3.3 shows the relationship between the length of a crack and the energy it takes to propagate that crack in a material such as glass. For cracks up to the *critical crack length, L_{crit},* more energy is required to create the two newly exposed surfaces (area) of the crack than is released by broken and relaxed chemical bonds in a volume of material around the crack. For cracks of greater than the critical crack length, more bond energy is being released by the growing crack than is being used to make new surfaces, so the crack propagates spontaneously. The reader is urged to read the spellbinding accounts by Gordon (1968 and 1978) on the importance of cracks and their propagation in Gothic cathedrals and ocean liners as well as in more familiar situations in our daily lives.

Because brittleness is a measure of how sensitive a material is to cracks, it can be expressed as the reciprocal of the critical crack length for a material. Brittleness is thus a material property that depends on the magnitude of a morphological flaw!

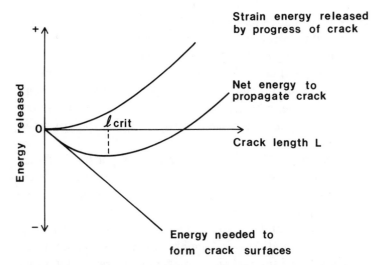

Figure 3.3 Graph showing the relationship between crack length and the energy required to make the crack grow. See explanation in text. Drawing by L. Blalock.

It also means that, in a batch of similar items (glass fibers, for example), the strongest item in the batch is the one with the smallest crack. As the batch is pulled on and all existing cracks grow, the longest crack will reach the critical length first, and its fiber will break, and the fiber with the shortest crack will fail last.

Although most brittle materials are rigid, some are soft and even stretchy. A rubber balloon has enormous breaking strain, but if at any time in blowing up the balloon you introduce a crack in it with a pin, the crack will propagate faster than your eye can follow it. The release of energy in the rubber happens so suddenly that you can hear it pop.

The critical crack length for brittle materials is very small— on the order of nanometers or micrometers. One way to decrease the probability that a piece of brittle material will crack is to reduce its diameter to less than the critical crack length; the boron whiskers that reinforce the plastic in helicopter rotor blades are 1 μm in diameter and the fundamental mineral crystals in bone are 2 nm thick. It requires a lot of energy to propagate cracks across these two very tough materials. One is heartened to note that glassy and rubbery materials and mineral crystals are not found in large pieces in organisms either.

Composite materials have two or more components of different stiffness. Fiberglass, which comprises thin glass fibers bound together with a softer plastic, makes a tough composite material that has some of the strength of glass, the flexibility of plastic, and very little tendency to fail by cracking. Wood, bone, and skin are tough because the stiff components (cellulose, calcium phosphate, and collagen, respectively) are fibers or crystals of smaller diameter. Thus, when such a fiber breaks (Figure 3.4), the crack runs across the fiber into the

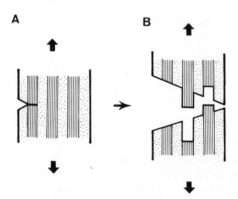

Figure 3.4 When a fiberglass rod is stretched, a crack that splits a glass fiber (A) stops because its energy is dissipated in the softer plastic. It takes a lot of energy to break a composite material because the crack must create much new surface area (B) as it goes around broken fibers. Drawings by L. Croner.

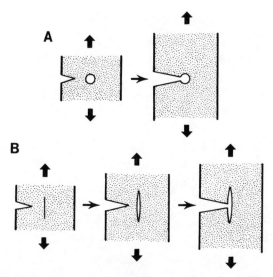

Figure 3.5 Cracks can be stopped by causing them to run into holes (A) or slits (B) because the stress required to make the crack grow is inversely related to the radius at the tip of the crack. Drawings by L. Croner.

pliant matrix and its energy is dispersed in all directions. The crack stops because the stress is too low to cause dangerous strains in the stretchier component.

As a crack propagates, it supplies the energy to create two newly exposed surfaces. Any deviation in the direction of the crack raises the energy required to drive it ahead, so the presence of reinforcing particles, fibers, or plates that increase the surface area a crack must create to go around them are effective crack-resisting devices.

Another way to stop cracks is to let them run into holes in the material (Figure 3.5). The stress required to propagate a crack is inversely related to the radius of curvature, R, of the tip of the crack. So, if a crack with $R = 1$ nm runs into a round hole with a radius of 1 mm, a million times more

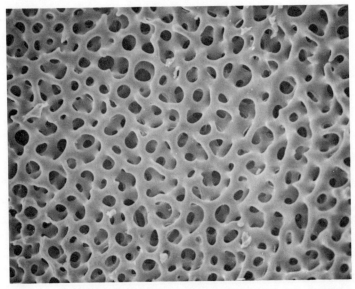

Figure 3.6 Scanning electron micrograph (SEM) of a brittle, calcified ossicle ("little bone") in the skeleton of a sea star. Its holeyness makes it a tough, crack-retarding material.

energy will be required to make the crack go further, and the crack may well stop. Bone material is full of holes, which contain cells and nerves and blood vessels. These rounded voids act as crack stoppers. The calcitic stereom of sea stars (Figure 3.6) and other echinoderms is perhaps the most striking example of a material whose intrinsic brittleness is ameliorated by being full of crack-stopping holes.

DEFORMABILITY AND WATER CONTENT

Although familiar biological composite materials—bone, shell, and wood—are rigid, most biocomposites are pliant.

Deformability of biological materials is ultimately due to the presence of liquid and, more specifically, to the low viscosity of water and fat or other oily substances in the materials. Every biomaterial has water in it, and many tissues are mostly water.

Fluids (both liquids and gases) are not elastic. That is, they do not resist being deformed. Instead, they are *viscous:* they resist the rate of deformation. Water readily flows to fit any shape of vessel we put it in, and when we move very slowly in water, we are unaware that anything is opposing our movement. On the other hand, we must push very hard to move rapidly through water. The water in biocomposites such as skin and cartilage causes them to respond more slowly to being stretched, twisted, and so on than do brittle materials. Pliant biocomposites are, in fact, among the toughest materials on Earth: they are hard to tear. Compare the bending behavior of a live twig with a dead twig from the same tree. The major difference between the two is that the live one has lots of water in it. The live one is tough and bends a lot before it breaks (high breaking strain), whereas the dead twig is both weak and brittle: it cracks in two with little strain.

Water in the material decreases the amount of elastic recoil the material will display. Water-damped elasticity is called *viscoelasticity* and retards both the deformation and the recoil processes. Inside the squashy material, frictional forces are created as the fluid moves past the solid components during deformation. This situation is found in the articular cartilage on the ends of human femurs and tibias (Myers and Mow, 1983), where viscoelasticity enhances shock absorption. At each joint in the human leg, a thin layer of articular cartilage on the end of the bone takes the full impact force of walking, running, and jumping. As the cartilage is squashed when the foot hits the ground, there is viscous flow of water sideways within the cartilage, away from the point of impact. This flow disperses and reduces the stresses, because it takes time and retards the rate of deformation.

Viscoelasticity is largely responsible for the nonlinear shapes of stress–strain curves (Figures 1.2 and 3.1). The stiffness (slope of the stress–strain curve) of a material such as skin and kelp stipe and even bone in its plastic region is nonlinear: stiffness depends on the stress being applied and on the amount of time and strain at the moment of interest. So to report a single value for the stiffness of a viscoelastic material is an oversimplification.

Another way to get a feel for the viscous aspects of soft biomaterials is to record the unloading curve on the same graph with the loading stress–strain curve (Figure 3.7) during a mechanical test of the material. Over most of the functional elastic part of the stress–strain curve of an elastic material such as bone or wood, the unloading curve lies so close to the loading curve that the graph shows a single line. A viscoelastic material (or wood and bone in their plastic range) will have a very different loading and unloading curve (Figure 3.7): at any given value of strain, the material has a greater stress during loading than during unloading. Energy has been lost as heat as a result of friction between molecules, and the total useful recoil energy is less than the energy it took to deform the material in the first place. The area between the curves is a measure of the inefficiency of the material's elastic recoil. The ratio of the area between the curves divided by the area under the loading curve is called *hysteresis* or the amount of elastic energy lost to the system in the loading–unloading cycle.

Elastic materials transmit most of the energy applied to them: they bounce or transmit muscular force and shortening to the skeleton during locomotion. Viscoelastic materials have a damping effect on loads applied to them. For example,

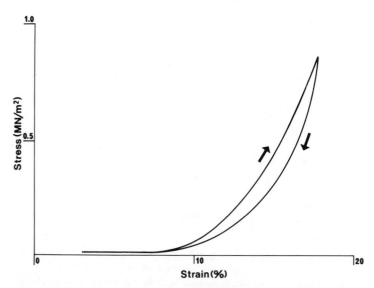

Figure 3.7 Stress–strain graph of whale blubber showing that the energy released when the load is removed (*descending arrow*) is less than the energy put into the material upon loading (*ascending arrow*). The area between the curves is the hysteresis: it represents the amount of energy lost to the system. Drawing by L. Srba.

our intervertebral disks and articular cartilages and the water-filled tissues of plants make good shock absorbers. Through hysteresis, they dissipate the high stresses due to impact that occur when we run on hard surfaces, and they prevent these forces from fracturing the rigid parts of the skeletal system.

In addition to viscous components, there are also elastic, rubberlike components in some biocomposites. The molecules of rubbery materials are very long and coiled and are cross-linked at infrequent intervals along their length. Thus, each molecule is relatively free to thrash about with random thermal motion. Figure 3.8 shows, very diagrammatically, that a randomly thrashing molecule of rubber occupies a

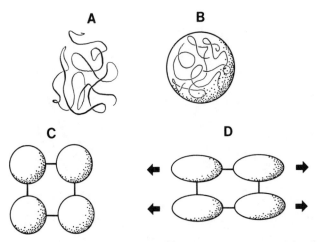

Figure 3.8 The basis of rubbery elasticity. (A) A rubber molecule is enormously long and is in constant, random thermal motion. (B) It thrashes about in a spherical domain. (C) In a piece of rubber, the molecules are linked together (shown diagrammatically). (D) When the rubber is stretched, each molecule is constrained to thrash more often in the direction of the force, and its domain is stretched into a football's shape. When the force is removed, the molecule and the piece of rubber immediately return to their original shape. See text for further explanation. Drawings by L. Croner.

sphere-shaped space. When the rubber is stretched, the spherical envelope of space is pulled into the shape of a football and the molecule spends relatively more of its time with part of its length parallel to the stress axis than it does in other directions. The thrashing activity tends to return the rubbery molecule to its spherical space, and it takes energy to constrain it to the longer, thinner envelope. As rubbery material resists this stretch, it generates heat. When the material is released, it returns very rapidly to its original shape and its spherical molecular envelope, and heat is lost.

You can actually feel this evidence of thermodynamic energy release: take a broad rubber band or a collapsed balloon

ening, thus allowing the colony to bend right over until it lies parallel to the flow. This mechanism reduces the surface area of the coral that is pushed on directly by the flow and so reduces the fluid forces on the coral. By stress softening and bending, the coral has avoided breaking. Muscles and an intake water pump in the coral colony can re-erect the coral after the flow has ceased.

In plant and animal bodies, fibers contribute most to achieving and maintaining the cylindrical body form. Collagen is the predominant fibrous material for animals of most phyla, and cellulose is the one found in cell walls of most plants. Chitin is found in some fungi and is a major structural component for a wide range of quite different skeletal materials in arthropods (Neville, 1975; Vincent, 1982), brachiopods, annelids, and mollusks. The property of cellulose and collagen fibers that is most important to their contribution to structure at the next level up the structural hierarchy is their enormous length to diameter ratio, by which they control the anisotropy of cords, membranes, and lumps according to their degree of preferred orientation.

Figure 3.11 shows stress–strain curves for membranes pulled at various angles with respect to the orientation of their reinforcing fibers. Tendons and plant fibers such as hemp and cotton have virtually perfect parallel fibrillar alignment and, when pulled on, they give the straightest, steepest stress–strain curves known for biopolymers (Figure 3.11A). Figure 3.11B shows, as one would expect, that materials reinforced with parallel fibers offer very little resistance to being pulled apart in the direction normal to the fiber axis.

The most interesting and important shape of stress–strain curve, and the one shown by most soft materials in plants and animals, is the J-shaped curve (Figure 3.11C). The J-shape is most often caused by an array of fibers that lie at various angles to the line of stress: during the stretch, the fibers be-

and, while holding it against your heat-sensitive upper lip, quickly stretch it. You can feel the heat generated by constraint of the molecular motions into nonspherical spaces. Now, holding it stretched out, wave it around to let it cool to room temperature, then hold it again against your lip and allow it to recoil: it cools noticeably. A kinesthetic appreciation of thermodynamics is most reassuring.

ANISOTROPY IN BIOMATERIALS

No matter which way you stretch or squash a solid rubber ball, the stiffness will always be the same. So will strength, toughness, and all other physical properties such as thermal expansion, electrical conductivity, and so on. We say that a material with each property having the same value in all directions is *isotropic*. Most structural biomaterials are *anisotropic;* that is, each physical property will have different values in different directions within the material. Wood, bone, and tendon are all much stiffer and stronger in the direction along their length (parallel to their "grain" or fibrous structure) than they are in other directions across the grain (Figure 3.9).

Anisotropy is caused by the structure of the material at the microscopic, ultrastructural, and molecular levels: the macromolecular components (fibers and crystals that are big enough to be seen in the light microscope) of the materials have a preferred orientation parallel to the long axis of the piece of material or skeletal element. Long, straight-chain protein and polysaccharide molecules oriented preferentially in one direction will give greater structural strength and integrity in the direction of the preferred molecular orientation. Strength and stiffness of materials are related causally to the strengths, densities, and angular orientations of chemical bonds.

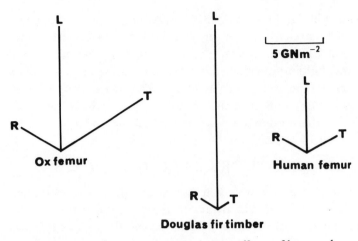

Figure 3.9 Diagram showing the relative stiffness of bone and wood in three orthogonal directions. *L,* longitudinal (parallel to the grain); *R,* radial; *T,* tangential. Drawings by L. Srba.

Although there is no rubber or rubberlike solid in a rubber tree (what is present is latex, which is a solution containing the monomer that we polymerize to make rubber), animals have tendons and ligaments that often contain rubberlike proteins. Elastin is the protein concentrated in the nuchal ligament, which helps to support the heads of quadrupedal mammals; and resilin is found in the flight systems of insects, where it serves to decelerate the wing at the end of a stroke and to accelerate it at the beginning of a stroke. Elastin is a regular component of many vertebrate connective tissues, and resilin is a regular component of both rigid and soft insect cuticular materials.

We add carbon to rubber to make it stiffer for use in tires; and soft corals and sponges respectively add crystalline grains of calcite and larger bits of calcite or glass that stiffen their polymeric structural materials (Koehl, 1982). Some sponges

make their own sclerites and some incorporate sar they pick up from the environment (Teragawa, 198

When material that is filled with particles is p stress, each filler particle causes matrix molecules t around it. Thus, the filler particles may contribute ness, but they also increase the likelihood that th molecules will reach their breaking strain. When fil rials are stretched a lot, breakage of matrix molecule loss of stiffness and a decrease in the strength of the which is accompanied by increased extensibility. T nomenon is called *stress softening*. Here is how it mi in a soft coral such as dead man's fingers, common Atlantic shores (Koehl, 1982). Imagine soft, fing branches of a coral colony in still water. As water f one direction bends the colony, the filler spicules coelastic skeletal tissue act initially to increase the sti to oppose bending (Figure 3.10). Faster flow or fl extended time may cause the material to undergo s

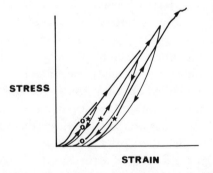

Figure 3.10 Stress–strain graph showing stress-softening tive tissue from the soft coral, dead-man's fingers (*Alcyoni tum*), in response to repeated loading and unloading. Stars the same stress causes increasingly larger extensions. Circle that a particular extension requires less and less stress on su loadings. Graph prepared by L. Srba.

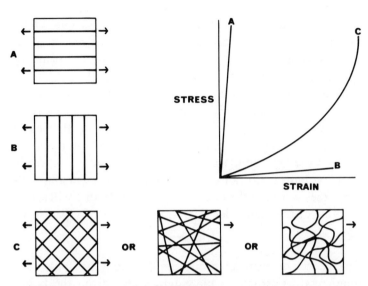

Figure 3.11 Stress–strain graphs of fiber-reinforced membranes. See text for details. Drawings by L. Srba.

come more and more parallel to the stress, and their stiffness adds increasingly to the stiffness of the material.

You can perceive the change in stiffness (indicated by the J-shaped curve) of a material by tugging at an ear lobe (or a pinch of skin on the inside of your forearm or the back of your hand). The first few millimeters of stretch come very easily; the next few millimeters are sensibly stiffer; beyond that the entire ear is being pulled downward and it hurts. This is a neat demonstration of the function of the J-shaped relationship between stress and strain as a safety factor. A potentially harmful tug is allowed some fairly easy stretch, and often the tug goes no further. Additional stretch is opposed but is still neither painful nor damaging. Finally, if the tug persists, the full stiffness of the material resists it, and a completely different mechanism, pain, tells you to take behavioral action to stop the tug before ripping begins. In spite of the fact that it has been taken for granted until recently, the usefulness of controlled stretchiness in animal and plant tissues cannot be overestimated.

Material with a J-curve stores much less energy than a linearly elastic material such as steel or glass, whose stress–strain graph is a straight line. J. E. Gordon (1978), a naval architect turned plastics expert, discovered the importance of the J-curve. He points out that for materials to be really stretchy and not to fail catastrophically like a bursting balloon, the J-curve is the only kind of elasticity that will serve. It is one of the many elements in the success of organisms.

The elastic mechanical properties (strength, stiffness, and toughness) of structural biomaterials arise from the bonds and other interactions within and between molecules in the materials. The predominant structural molecules of biomaterials are polymers of high molecular weight: proteins and polysaccharides. Water and oil or fat, either bound to the molecules or flowing around them, determine the materials' time-dependent, viscous properties and hysteresis. The preferred orientation of collagenous, chitinous, cellulosic fibers, and arrays of mineral crystals cause materials such as wood, bone, shell, tendon, and arthropod cuticle to be anisotropic. Voids and the continuous matrix of composite materials increase toughness by decreasing stress concentrations and the chance of brittle fracture. Loosely linked, randomly coiled, and thermally thrashing molecules such as elastin and resilin give materials rubberlike properties. The reinforcement of membranes with stiff fibers in biaxial or random arrays gives them a J-shaped stress–strain curve: a remarkable shape control mechanism and safety factor of soft, stretchy tissues.

4

STRUCTURAL SYSTEMS
OF SHAPED ELEMENTS

Just as the structure of biomaterials governs their func-
tional properties, so too does the construction of whole
systems permit, control, and limit complex, higher level
functions such as feeding, fighting, and locomotion. A structural system may be an entire organism or it may be a part,
such as a tentacle or a stamen. It is at the system level that the
compatibility of materials mentioned in Chapter 1 is especially important: the stiffness, stretchiness, and viscoelasticity
of adjacent materials must be in harmony with the forces to
be sustained and the motions to be made in the time appropriate to the action.

Although we will identify three types of systems—
branched cylinders, hydrostats, and kinetic frameworks—
these are arbitrary categories, and there are no distinct
boundaries between the types. Many animal bodies, such as
our own, are dizzy mixtures of all three types. But what all
organisms have in common is a cylindrical shape of the body,
roots, branches, appendages, and internal conduits.

Most plants, corals, hydroids, and bryozoans are branched
cylinders. There is smooth continuity of material across the
joints, which are no more flexible than the stems on either

side. Hydrostatic bodies are containers of watery fluid. In sea anemones and some large worms, the fluid occupies a single, large body cavity. Most tiny worms, tongues, and elephants' trunks keep the water in single cells, but they operate by the same rules that permit and constrain hollow hydrostats. Kinetic frameworks may be either wormlike bodies, such as fishes and snakes, or branched cylindrical bodies like our own and those of insects and sea stars. Discontinuity of materials at our joints renders us flexible.

Animals have muscles that move their appendages and bodies by creating forces that are applied to the mechanical support system. According to the mechanical properties of the skeletal elements and the geometry of their arrangement, the forces or the motions and speed of muscular activity may be magnified. The multiplication of force by a complex system of force-producing elements and passive support elements is called mechanical advantage. For levers (Figure 4.1), the force applied, F_{in}, times the length of its lever arm, L_{in}, equals the force output F_{out}, times the length of its lever arm, L_{out}.

$$F_{in} \times L_{in} = F_{out} \times L_{out}$$

Mechanical advantage is the ratio of the length of the input lever arm to the length of the output lever arm: L_{in}/L_{out}. It is the factor by which the lever multiplies the force applied by the muscle; and it may be greater or less than one.

$$F_{out} = F_{in} \times (L_{in}/L_{out})$$

The force output is therefore equal to the applied force times the mechanical advantage.

Figure 4.1 also shows that the distance moved by either end of a lever is proportional to the length of its lever arm. When a lever rocks about its fulcrum, both ends move their respective distances at the same time. The end of the longer lever

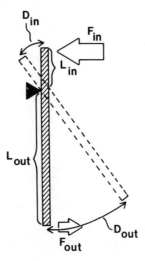

Figure 4.1 Leverage system. See text for explanation. Drawing by L. Blalock.

arm will move a greater distance at a greater speed (speed = distance/time) than will the end of the shorter lever arm. A longer output lever arm will therefore have a higher *displacement advantage* and *speed advantage* (L_{out}/L_{in}), but a lower mechanical advantage, than a short one will.

These "advantages" have been useful in analyzing force and power outputs of movements of stiff-legged creatures like arthropods and vertebrates. But until Kier and Smith (1985) showed the significance of the curve in Figure 4.2 and the relationship

$$\text{volume of a cylinder} = \pi r^2 L,$$

it was not realized that soft-bodied animals and muscular organs like tongues and tentacles, which have no rigid levers, can also have mechanical advantage, displacement, and speed advantages. These advantages in soft-bodied animals are

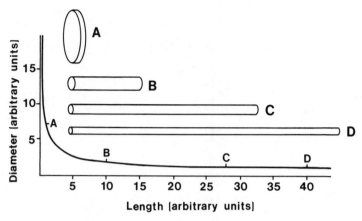

Figure 4.2 Graph showing the relation between diameter and length of a cylinder of constant volume. Drawing by L. Blalock.

made possible by the constant volume of the cylinder and the r^2 term in the expression for volume of the cylindrical body form. Figure 4.2 shows that the original shape of the cylinder (specifically, its length-to-diameter ratio) controls which muscle set has which advantage. Thus, in a long, thin cylinder of constant volume, a small decrease in radius effected by transverse or circumferential muscles will be accompanied by a large and rapid increase in length of low force. In a short, fat cylinder, a small decrease in length, created by contraction of longitudinal muscles, causes a moderate but forceful increase in diameter.

A lively example is the lightning-fast extension of the squid's tentacle. Before its attack on a shrimp, it has a length-to-diameter ratio of 10:1. If we find this point on the graph, we see that a tiny decrease in the diameter of the cylinder will result in an enormous increase in length. Kier (1983) showed that the small squid *Loliguncula brevis* extends its tentacles by 70 percent in 30 ms in catching the shrimp they eat. In the high-speed films of this action, the change in diameter of the tentacles that produces this enormous elongation is almost too small to measure!

The earthworm is more subtle. It uses circumferential muscles in its front end to extend ahead into the soil as it burrows. It is segmented and therefore not a single hydrostat. Instead its body is a long string of large, fluid-filled hydrostatic cavities, one per segment. When the earthworm extends its body out of the soil and noses around looking for a mate, the last hydrostatic segments remain in the burrow. If a bird threatens to pull the worm out of the burrow, the worm has a system of giant nerve fibers that tell all the longitudinal muscles in the body to contract at once and pull the worm back into the burrow, thereby enabling the worm to escape. This action depends on the anchoring action of the tail of the worm, which presses outward against the side of the burrow. A cylinder within a tube can expand and apply high pressure against the tube and thus create high friction, which would anchor the cylinder in the tube. The earthworm is taking advantage of the relation shown in the upper end of the swooping curve in Figure 4.2, because the segment in the ground has a low length-to-diameter ratio. Thus, a small, longitudinal contraction will expand the worm forcefully against its burrow and anchor it tightly.

As we survey organismic structural systems, it is instructive to keep in mind the difference between the processes by which biological structures develop and the manner in which man-made ones are created. The biological process of starting as a tiny blob and growing and developing into a larger but still embryonic structure, then into an even larger juvenile one, and finally into the fully formed adult is simply not practical technologically or, for that matter desirable, but we plants and animals do it every day. Young, small stages of organisms often have functions different from those of the adult; for example, mobile larvae of immobile, bottom-dwelling aquatic creatures are responsible for the dispersal of the species into new habitats. This is not the way we build most things: although some houses are indeed built one section at a time, a couch is not made by elongating a chair and a truck is not a grown-up car. Your tricycle does not grow up with you and become a bicycle.

BRANCHED CYLINDERS

Plants, coral and bryozoan colonies, and some sponges are simply systems of branched cylinders (Figure 4.3). Even

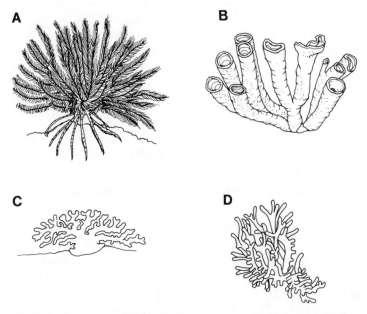

Figure 4.3 Branched cylindrical organisms. (A) Feather star (Echinodermata); (B) sponge; (C) hydrocoral (Cnidaria); (D) calcareous alga. Drawings by L. Srba.

though the biological functions of all branches of an individual organism differ, stem from root, arm from leg, the principles that direct or explain their mechanical design are the same. For example, in palm trees, the trunk, the fronds, and the stem of the flowers have different diameters, materials (tissues), and cross-sectional shapes; and they bear different structures with different functions. However, they are all beams whose cross-sectional shapes and distribution of materials of different stiffness are appropriate for their support.

The crucial parts of branched cylinder systems are the branch points. Human designers are challenged by the fastening of one part to another in cars, shirts, and clarinets. Even when the material is the same on both sides of the joint, as in a wooden chair, the creation of exact fit and the right application of screws or glue takes thought, care, and skill. The best evidence that joints are difficult to design and build is that failure so often occurs first at the joints.

Now compare the chair with a tree and note the smooth continuity of wood and bark from trunk to branch: there is no cement, welding, nails, or dovetailed joint. Instead, the branch grows from a bud, and continuous growth of wood surrounding the entire structure produces a designer's dream, which we can achieve only with the technology and high heat used in blowing glass and casting metals, plastics, and fired ceramics. Spliced rope and woven fabrics make excellent tensile joints, but the brittleness and low tensile strength of unfired clay render it nearly useless mechanically.

The advantage of the branched, cylindrical body lies in the smooth transmission of forces across the joints. The uninterrupted structural transition from trunk to branch assures that when a load is applied to the tip of a branch, the force in the branch is transmitted evenly and over a large area to the trunk. Imagine attaching the branches in any other way. In most man-made structures, structural elements (branches) are made separately and then attached to each other, usually with nails or screws or brackets or glue. Unless the builder is experienced, the joints, being made of materials different from those of the framework, will tend to concentrate stresses, and the branches will break away at these disharmonious joints, or the builder may overbuild the joints to avoid this problem and thus create another—the wasteful use of material.

All plants and animals with the branched-cylinder body plan are sessile; that is, they are attached to some solid substrate and extend their branches into the ambient air or water and their roots or holdfasts into the substrate. Corals, sponges, and large terrestrial plants are fairly rigid branching systems. Most plants, including the twigs of the stiffest trees, are flexible enough to sway with the wind or waves, but stony corals and many sponges are quite inflexible. Some branched-cylinder organisms do have joints of greater flexibility, and the secrets to the integrity of the body are still the continuity of the formative tissue and its production of new material. These joints are flexible because they have a small diameter (smaller I) or are composed of a more compliant material (lower E), or both. The branched-cylinder system works for complex and even for very large structures (such as giant sequoia trees) and is the envy of engineers, architects, and builders. Indeed, it is a sensible design for complex bodies.

HYDROSTATS

An example of a *hollow, hydrostatic skeletal system* is the turgid balloon (Figure 4.4). It is a system of two parts, both of which are necessary to the supportive function. The rubber is in tension because it is stretched around and compresses the contents of the balloon. And the compressed volume (not the

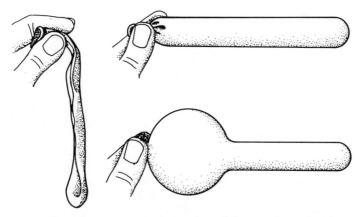

Figure 4.4 A hydrostatic skeletal system such as a pressurized balloon (*top right*) has a tension-resisting membrane surrounding a compression-resisting volume of fluid. Without pressure (*left*) there is no support. An increase in pressure in a cylinder (*bottom right*) produces an aneurysm. Drawings by L. DeLeon.

shape) of the air or water in the balloon produces tension that stretches the rubber membrane. Some organisms such as sea anemones, nematode worms, caterpillars, vertebrates, and the tube feet of echinoderms (Figure 4.5) really are great fluid-filled cavities surrounded by a body wall, just like water-filled balloons.

Many other animals have *muscular hydrostatic systems*. In these systems water is packaged in many muscle cells instead of in one great cavity: membrane-bound cells are the tensile, elastic containers. Most molluskan bodies, the tentacles and arms of squid and octopus, our lips and tongue, and the elephant's trunk are examples of muscular hydrostats (Figure 4.6). These flexible organs have longitudinal muscles that cause shortening or bending. To produce a bend, the longitudinal muscles on one side contract at the same time that circumferential or transverse muscles contract on the opposite

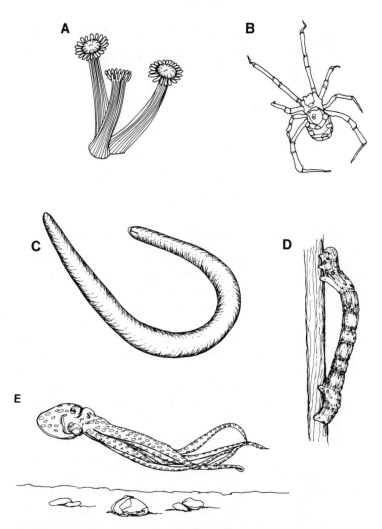

Figure 4.5 Animals with large hydrostatic body cavities. (A) Coral polyps; (B) spider; (C) nematode worm; (D) inchworm caterpillar of a moth. (E) The octopus swims by expelling water from its hydrodynamic gill cavity, but its tentacles are solid, muscular hydrostats. Drawings by L. Srba.

side, normal to the long axis (Kier and Smith, 1985). When longitudinal muscles all contract at once, of course, they cause the organ to shorten. Extension of the organ is accomplished by the contraction of circumferential, radial, or transverse muscles—any muscle that will cause a decrease in the diameter of the cylindrical organ.

Pressure in a fluid-filled vessel pushes out with equal force in all directions. In a spherical vessel, the stress in the wall will be the same everywhere and in all directions in the wall. But any lengthening of one axis in the sphere—a change

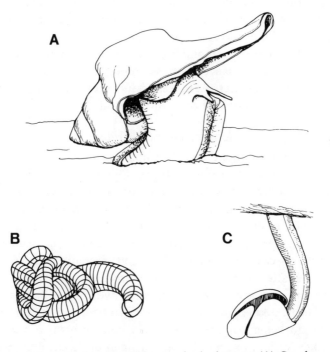

Figure 4.6 Animals with solid muscular hydrostats. (A) Conch; (B) nemertine worm; (C) stalked barnacle. In the barnacle, the stalk is the muscular hydrostat. Drawings by L. Srba.

producing a cylindrical shape—gives rise to mechanical anisotropy: *In the wall of a pressurized cylinder, the stress, S_C, exerted by the contents on the circumference is twice the stress, S_L, exerted on the length: $S_C = 2S_L$.* In other words, because the body shape is cylindrical, the stresses in its wall due to pressure inside are not equal: the pressure-generated stress in a cylinder tending to burst its belt is twice the stress tending to blow off the ends. Thus, when you start to inflate a cylindrical balloon, a bulge appears (Figure 4.4). The bulge is visual evidence of this mechanical principle: the balloon is getting fatter rather than longer. This same principle explains the formation of an aneurysm in a cerebral artery. Unbalanced stresses in the walls present a problem in the design of pressurized cylinders (vessels and pipes, gas and water tanks, swim bladders of fish, and lungs), whether they be man's or nature's. An imbalance of stresses in different directions in a homogeneous material can lead to inappropriate bulging and to explosive rupture of the cylinder wall.

Reinforcement by Fibers

Engineers have found that the imbalance of stresses in cylinder walls can be corrected by wrapping the cylinder in a helical pattern with stiff fibers. Figure 4.7 (left) shows that if the fibers are wrapped circumferentially around a cylinder and their position stabilized by longitudinal fibers, there will be no danger of bulging. But there also can be no change in length, because the longitudinal and circumferential fibers will not stretch in their respective directions. A worse problem with the orthogonally reinforced hydrostat is that, when it is bent, it kinks, because longitudinal fibers on the concave side are being compressed—when you push on a rope, it kinks. A kink, like a bulge, is a dangerous mechanical instability. If a bending cylindrical body—for example, a swimming fish—kinks, bending will suddenly accelerate, and the

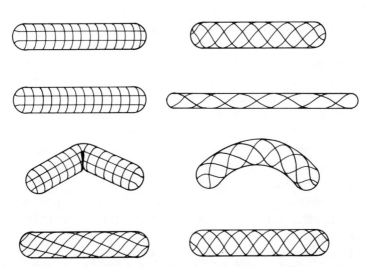

Figure 4.7 Fiber-reinforced, pressurized cylinders. On the cylinders on the left, the fibers are oriented longitudinally and circumferentially. Although this prevents bulging, it also prevents changes in length, and it causes the cylinder to kink during bending. It also allows torsion. On the cylinders on the right, the fibers are wrapped helically around the cylinder. This prevents bulging and torsion, but allows changes in length and, most important, bending without kinking. Drawings by L. DeLeon.

body will achieve a totally inappropriate, nonstreamlined shape. Kinks are to be avoided in pressurized cylinders. Helically wound fibers (Figure 4.7, right) solve all these problems at once: bulging is avoided and smooth bends are permitted. One can say that helically wound, pressurized cylinders are constrained to bend smoothly without kinking.

Although many kinds of animals are hollow, one never sees earthworms or caterpillars with aneurysms. Our blood vessels normally show no such deformations, and they function by accommodating pressure in the fluids they contain. Hollow hydrostatic cylinders in living animals and in the muscles in some animal hydrostats have a structural feature that counteracts the imbalance of directional stresses: the cylinders are wrapped helically with stiff fibers, usually of collagen, in their wall material. Because we find stiff, helically wound fibers around pressurized hydrostats in plants and animals, we presume that the structural feature of stiff, helically wound fibers is always related to the hydrostatic function. Indeed, in the systems that have been studied (sea anemones, nematodes, nemertines, squid and octopus appendages, fishes, and whales), the helically wound structure apparently functions mechanically in the posture and motion of its host. We can predict that wherever we find the helical fiber systems wound around cylindrical bodies, we can expect to find high internal pressure, or the ability to bend the body without causing wrinkles, or both. Alternatively, we can predict that cylindrical, muscular bodies that bend without wrinkles, or those that are found to have high internal pressure, will have a reinforcing array of stiff fibers wrapped helically around the body.

Interestingly, the flexible arms of sea stars are hollow, cylindrical, and fluid-filled, but the array of stiff collagen fibers in the walls of the arms is not helical. Instead, it is an orthogonal array with fibers coursing circumferentially around the arm and longitudinally along it (P. O'Neil, personal communication). This is not the exception that disproves the rule. Rather, it is consistent with our rule because the fluid in the arm is not under high enough pressure to be affected by helical fiber winding when the arm bends. Also, a sea star rights itself from an upside-down position by twisting one or two arms so the tube feet can grab the substratum (Figure 4.8). Twisting of low-pressure cylinders is allowed by orthogonal nets of stiff fibers in the wall but resisted by helical arrays (Figure 4.7, left).

The combination of the imbalance of stresses and the pres-

A

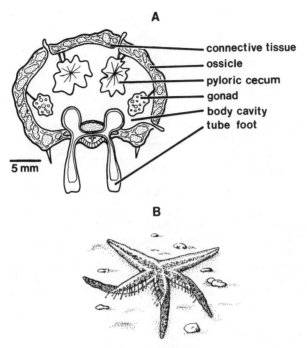

connective tissue
ossicle
pyloric cecum
gonad
body cavity
tube foot

5 mm

B

Figure 4.8 Sea star organization. (A) Cross section of an arm, showing the large, low-pressure, fluid-filled body cavity and two of the compliant, hydrostatic tube feet. (B) When a sea star is turned onto its back, it rights itself by twisting an arm around until the tube feet can attach to the substrate and pull the animal upright. Drawings by L. Croner.

ence of an anisotropic, fiber-reinforced membrane provides hydrostatic cylinders with some interesting functional properties that allow for quite a diverse range of behavioral capabilities among organisms. The *fiber angle* is the angle made by a helical fiber and the long axis of the cylinder it wraps. Pressurized cylindrical bodies have two dramatically distinct behavioral capabilities depending on whether the fiber angles of crossed-helical arrays of stiff, reinforcing collagen fibers in

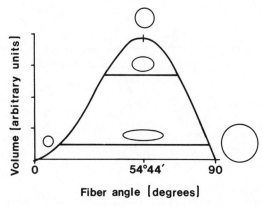

Figure 4.9 Graph showing how the volume of an open cylinder varies with the fiber angle of helically wrapped reinforcing fibers. Maximum volume is contained in such a cylinder when the fiber angle is 54° 44'. Circles indicate that the cylinder is thin (and long) at low fiber angles and fat (but short) at high fiber angles. Any horizontal line describes the limits of short-and-fat or long-and-thin that a constant volume cylinder may have: the flatter the elliptical shape (*ellipses*) the constant volume cylinder has when its fiber angle is 54° 44', the lower on the graph will be its horizontal line. Graph prepared by L. Blalock.

their walls are greater or less than 54° 44' (Figure 4.9) (Clark and Cowey, 1958; Clark, 1964; Wainwright et al., 1976). If the fiber angle is high, as it is in nematodes, sharks, and whales, then the changes in body curvature controlled by longitudinal muscles dominate the behavior. When the muscle on one side of the body contracts and bends the body to that side, the increased underlying pressure stretches the skin and its fiber array stretches, stiffly but elastically, a few percent. When the muscle relaxes, the elastic energy stored in the fiber array causes the body to recoil to its resting shape. Notice that no antagonistic muscles are necessary for this return to resting shape: the longitudinal muscles are working against the incompressible, constant volume of the hydrostat.

We had come to expect that all worms should have circumferential muscles as antagonists to longitudinal muscles: nematodes surprised us by lacking them completely.

In pressurized cylinders with low fiber angles, for example, squid mantles, huge circumferential muscles contract and change the body's diameter. Once again, the muscle contraction causes an increase in pressure in the underlying tissue that pushes out on the fiber array, forcefully stretching it a few percent. And again, when the muscle relaxes, the elastic energy stored in the collagen fiber array recoils and causes the mantle to regain its larger resting diameter. In each of these two designs, the passive elastic stiffness of the collagen fiber array is antagonistic to the action of the muscles: it stores enough elastic energy in the body wall to return the body to its resting shape after the muscle has relaxed.

The interaction of hydrostatic cylinders with their reinforcing, crossed, helically wound fibers depends absolutely and totally on the pressure of the underlying cavity, in hollow polyps and worms, in the body muscles in fishes and whales, or in the tongues, tentacles, and trunks of various animals (Kier and Smith, 1985; Wainwright et al., 1978). In other words, the cylindrical body form is uniquely able to use a previously unheralded and ignored function of contracting muscle, namely, the forceful increase in diameter that necessarily accompanies the well-known forceful contraction. Muscles can, and indeed do, push, and Nature has designed the cylindrical body plan of some of her liveliest creatures in a way that renders this action useful.

The thick walls of woody xylem cells that provide the support for woody plants are cylindrical. The cellulose molecules in the thick secondary layer of their walls are typically wound at an angle of about 20° around the cylindrical cells (Mark, 1967). The low cellulose fiber angle means that the woody tissue will rigidly and elastically resist longitudinal stress in the stem, such as that applied by the weight of the tree to the wood in the trunk.

There are collagen fibers in helical array in the walls of our major arteries. Do they function in the pumping of blood? We do not know. But this query is an example of a potentially important functional question that can arise from the application of a physical principle and an engineering equation to the analysis of function in a known but poorly understood biological structure.

There is scarcely a plant or an animal outside the phylum arthropoda that does not have a helical array of cellulose, chitin, or collagen fibers wrapped around some cylindrical structure in its body. If living bodies and their parts are generally cylindrical in shape, then the additional structural feature of helically wound reinforcing fibers in their outermost tissues makes a lot of sense as a means of ensuring the flexibility and regulating the stiffness and postural possibilities of the system.

Muscular Hydrostats

These soft, flexible systems have the same basic capabilities of permitting, controlling, and limiting postural changes possessed by hollow systems, but the soft, flexible systems have the advantage of not being liable to incapacitating punctures. Examples of muscular hydrostats include the solid tentacles of cephalopod mollusks and some hydroids, the crawling foot of snails, the probing foot of clams, and the tongues and proboscides of mollusks and vertebrates in general—especially fly-catching frogs and chameleons, taste-testing snakes and lizards, and manipulating tapirs and elephants.

The graph in Figure 4.2 shows the relationship between the length and diameter of a constant-volume cylinder. The swooping curve of this graph indicates that, depending on the shape the cylinder has before deformation, a given change in one dimension may cause the orthogonal dimension to change a little very forcefully or a lot less forcefully. Depending on their design, muscular hydrostats may be capable of mechanical or distance and speed advantages.

When a shark swims, muscles on one side of the body contract, thereby causing the body to bend and the tail to wag and create thrust that pushes the shark forward. The skin of the shark contains layers of parallel collagen arrays wrapped in right- and left-handed helices around the body. When muscle contracts on one side, it bulges, causing an increase in collagen fiber angle that helps to forcefully shorten the skin and its enclosed muscle on that side of the fish. The shark's cylindrical shape allows it to use the swelling of its contracting muscle to help it make forceful body-bending locomotor movements.

Although most hydrostats are cylindrical, they are functionally diverse. This functional diversity is related in large part to the ratio of the length to the diameter of the cylinder and the angle at which stiff, reinforcing fibers are wrapped around the cylindrical bodies. Two-dimensional bias-fabric membranes have some mechanical properties (for example, the J-shaped stress–strain curve) that one-dimensional fibers do not have. When the two-dimensional, bias-fabric membrane is rolled into a cylindrical three-dimensional body, the body has properties (for example, it bends without kinking and resists torsion) that two-dimensional membranes do not have. Thus a single feature (fibrous collagen) may take part in a different function at every level of the structural hierarchy.

KINETIC FRAMEWORKS

The supportive systems of arthropods, vertebrates, and some echinoderms are made up of rigid elements held together at flexible joints by soft polymeric materials. In vertebrates, echinoderms, and large crustaceans, the rigid, often cylindrical elements are mineralized. In arthropods other than crustaceans, they are made of a unique polymeric composite whose major components are chitin and proteins. Such frameworks of struts, ties, and cantilevers are diverse in their details, but they have in common the ability to use muscles of large cross-sectional area that can apply large forces as point forces to the rigid skeletal elements via tendons. The rigid elements are therefore useful as levers; and the resultant mechanical, displacement, and speed advantages are design features that adapt creatures to various modes of life. In other words, cylindrical rigid elements operated as levers by forceful muscles via tendons really make it possible to have paddles, wings, and legs, which can be either short and forceful or long and useful in gliding, in fast running, or in swinging from branch to branch.

Complex polychaete sea worms and earthworms have bilaterally symmetrical paired appendages that help them to burrow more forcefully or to swim faster. The arthropods with their exoskeletal kinetic frameworks and leveraging, paired appendages evolved from simple, hydrostatic burrowing worms. Fishes have unpaired dorsal, anal, and tail fins that increase the thrust-producing area of the bending cylindrical body, whereas their bilaterally paired fins are useful for maneuvering. Paired fins evolved into lever systems that aid amphibians, reptiles, birds, and mammals in burrowing, walking, running, and flying. Sea stars, sea urchins, sea cucumbers, sea lilies, and brittle stars (echinoderms) display a body design different from that of arthropods and vertebrates: their appendages are in cycles of five that are often arranged radially around the bilaterally organized body. Some of these appendages are hollow hydrostats and others are kinetic frameworks of flexible ligaments and rigid segments.

The basic design of the structural component of complex arthropods, vertebrates, and some echinoderms is a flexible, hollow or solid cylinder that resists changes in length. The cylinder is flexible because it is composed of alternating rigid and flexible segments. The long legs of arthropods and vertebrates and the mobile spines of sea urchins have few, long, levering limb segments, whereas articulated echinoderms, backbones, and multisegmented arthropod bodies trade the simple rigidity of tentpoles for the flexibility of many, short limb segments.

What is the possible advantage of making a supportive system of short, lumplike rigid elements? Our best guess is that the ancestral bodies of arthropods and vertebrates were soft and flexibly motile and that rigid elements evolved first as small, stiff lumps in the soft, polymeric connective tissues. Stiff lumps can accept point loads as part of axial or circumferential support. Once point loads are feasible, it is also feasible to have the enormous force of a very large muscle concentrated at one point via a tendon. This, in turn, is a condition for the most efficient use of levers. It appears that stiff skeletal rods, tendons, and large striated muscles capable of rapid contractions all evolved together in arthropods and vertebrates.

Very few attempts have been made to analyze the entire postural and locomotor support systems characteristic of particular animals. The parts we do see—and the structural connections and functional relationships between and among the parts—are apparently effective designs for doing what they do. A few species of burrowing, crawling, swimming, running, and flying animals have been analyzed as locomotor systems, and it appears that the single structural feature underlying the mechanical success of all these systems is their cylindrical shape.

The mechanical forces on Earth and their deformative effects on bodies follow well-known physical laws and therefore are predictable. Plants and animals that passively accommodate these forces contain materials that range from very pliable mucous layers to rigid bone, shell, and wood. The distribution of these materials in cylindrical supportive elements can be explained by mechanical design principles, such as beam theory, that have been derived from physical laws. The cylindrical body shape is mechanically the best shape to enable anchored organisms to grow into new space and to compete for wind- and waterborne resources. Cylindrical body shape is also best for streamlined swimmers and fliers and for stiff, levering appendages that promote fast or forceful locomotion on solid ground or in water or air. There are three distinct designs that account for the way cylindrical plants and animals maintain posture and move around: (1) the rather stiff, branched cylinder with continuous material across its joints, (2) the fiber-wound, fluid-filled hydrostat of wilting plants, wriggling worms, squirting squids, and twirling tongues, and (3) the kinetic framework with flexible joints that blesses brittle stars, crinoids, arthropods, and vertebrates.

5

THE ORIGIN OF THE CYLINDRICAL BODY SHAPE

Over evolutionary time, organisms have, in general, become larger, and the hierarchical structure of their bodies has become more complex. The increased complexity of the extracellular products of the single-celled organism may have conferred advantageous design features that led to the inevitable evolution of the multicellular cylindrical body form. Mechanics probably played a major role in the origin of form in plants and animals. Consider the primordial unicellular organism in the primeval ooze that appears in other fanciful accounts of early life on Earth. Ooze, it is presumed, was aqueous and viscous. Gravity, wind, and waves pushed bodies around according to the same physical laws that operate now. Assume that new structural features appeared sequentially, as though arising by mutations. Each structural feature conferred on the newly evolved form a functional attribute that its predecessor did not have. The addition of only four features could have produced a multicellular organism with a cylindrical body form. And the interaction of these features with the materials and structural units described in Chapters 2–4 can account for the shapes of all fossil and extant multicellular forms.

FEATURE 1: EXTRACELLULAR POLYMERS

Highly hydrated polysaccharides, proteins, and their mixtures—glycoproteins and proteoglycans—are produced by living cells today in various concentrations. These polymers form mucoid coats around cells (Figure 5.1A).

Attribute: Intercellular Adhesion   Mitotic siblings can adhere to each other and form a multicellular, shapeless blob (Figure 5.1B). Depending on their molecular structure, including the active side groups, extracellular polymers can confer a number of additional advantages on unicells, for example, protection from desiccation or noxious chemicals, adhesion to solid substrata, and mechanisms for the recognition and trapping of food particles or sexual partners.

The polysaccharide and protein material has the potential, like mucus, to be either a lubricant or an adhesive (Gosline and Denny, 1980). If the polymer sticks to itself more than to other objects, its surface tension will cause the multicellular

A   B

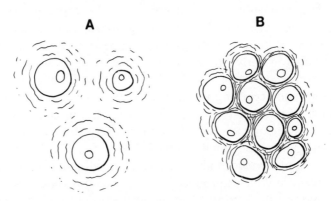

Figure 5.1   (A) Unicells shedding extracellular polymers into the primeval ooze. (B) First evolutionary step: change in the charge distribution on extracellular polymers causes daughter cells of mitosis to stick together. Drawings by L. Croner.

body to tend to round up into a sphere. If the polymer adheres more tightly to solid substrata, the blob may spread out in a thin, circular, platelike shape that takes the form of the substrate.

Note that one structural feature (the polymer) can give rise to another structural feature (multicellularity) at a higher level of organization via a functional property (adhesivity). Similarly, the polymer can produce a spherical or platelike shape by its selective adhesivity. Throughout this chapter, the concepts that (1) structure *permits* function and (2) changing structure *is* function are repeatedly exemplified.

FEATURE 2: CROSS-LINKING OF THE POLYMERS

Charged groups on polymers either attract and interact with one another in some type of bond or they repel each other. Cross-linking between polymers may be a matter of the formation of a covalent bond, a hydrogen bond, or other attractive interaction between two molecules or two sites on the same molecule. It can also refer to a molecule of one type, either small or large, being the link between two other large molecules. Figure 5.2 shows some of these possibilities.

Mucus itself varies among species and even from tissue to tissue in plants and animals. Figure 5.3 shows a mucus molecule as looking something like a long, limp bottle brush: it has a protein core and bristles of polysaccharides, which are covered with negative charges that repel each other and thus keep the bottle-brush shape expanded in water. Mucus molecules may be linked either to each other or to other substrata, for example, the cell surface. The sulfhydryl (-SH) groups shown on the naked end of the protein core function in such connections.

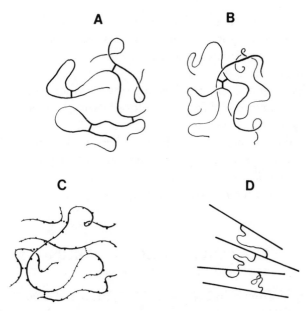

Figure 5.2 Cross-linking of polymers. (A) Strong (covalent or disulfide) bonds between or within polymer molecules. (B) Trivalent bonds. (C) Weak (hydrogen) bonds that can easily slip from site to site along the polymer. (D) Stiff, straight-chain polymers linked by long, flexible polymers. Drawings by L. Croner.

Attribute: Enormous Range of Material Stiffness
The particular stiffness of interest here is *shear stiffness*. You experience this property when you cause contiguous parts of a body to slide relative to each other (Figure 5.4). You can shear skin tissues by moving your thumb and forefinger against each other on either side of your earlobe or any pinch of skin. By scuffing your shoe on the floor, you shear off a thin layer of the shoe sole. The measure of shear stiffness is shear modulus, G, which is defined as τ/γ. The variable τ is the shear stress (force/area) and γ is the shear strain. Shear is an action in which real or imaginary layers throughout the

thickness of the material slide past one another; shear strain is the angle, measured in radians, through which a material has been displaced (Figure 5.4). Shear modulus is analogous to the tensile elastic modulus, *E,* discussed in Chapter 3, because it expresses the amount of force it takes to produce a unit of shearing deformation.

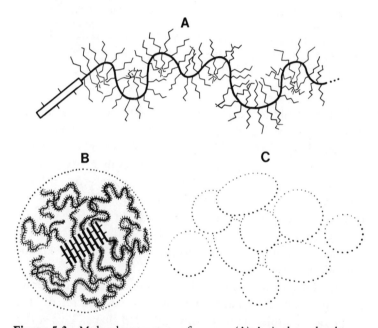

Figure 5.3 Molecular structure of mucus. (A) A single molecule showing the long, flexible protein core with negatively charged polysaccharide side groups that repel each other and give the molecule a shape like a bottle brush. One end has no large side groups, but it does have sulfhydryl groups that can interact with similar groups on other molecules, as shown in (B). The dotted circle in B represents the domain occupied by this cross-linked group of molecules. (C) Slippery mucus is a slurry of such domains, which are free to slide over one another. When mucus is stiff and elastic and acts as an adhesive, the molecular groups are linked to each other. Drawings by L. Croner.

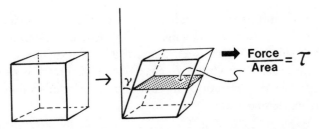

Figure 5.4 Shearing. When a block of solid or fluid is sheared, one face is moved relative to the opposite parallel face. Shear stress (τ) is the force/area parallel to the force. Shear strain (γ) is the angle through which the perpendicular faces of the block have been rotated. Drawing by L. Croner.

Cells control the shear modulus of their intercellular matrix in several ways. Figure 5.5A shows a simple linear polymer with no charged groups. The molecule is suspended in water and is thrashing about randomly as a result of its thermal energy. It is highly deformable; at every nanosecond it has a different shape. Although it may become entangled, it attaches to no other molecule and contributes only the volume of its envelope of thrashing to the shear stiffness of the slurry it makes with others like it. This particular molecule could be a good fluid lubricant.

The molecule in Figure 5.5B has anionic side groups along its length. The like charges repel each other and restrict the shape of the molecule to a more unfolded shape. This stiffer molecule contributes its increased stiffness to the matrix material. Bonds between a charged polymer and other molecules (Figure 5.5C) are perhaps the most important means of increasing matrix stiffness in polymeric biomaterials. Increased bonding leads to progressively stiffer and less stretchy matrix materials. But if the links between stiff molecules are themselves long and mobile, then the material can be a stretchy solid—an extensible polymeric network. Any process that restricts the random thermal motion of polymers is a shear-stiffening mechanism.

Solutes and solvents both affect the conformation and stiffness of polymers. Many extracellular polymers have high densities of anionic charges that can interact with mono- or divalent cations in the surrounding fluid. Monovalent cations shield the polymeric anions, thus allowing tighter conformation, whereas divalent cations form high-energy bonds between anionic sites on the same or adjacent molecules (Figure 5.5D). Nonpolar side groups (those that do not ionize) prefer to react with other nonpolar groups and to resist interaction with polar groups. These interactions also affect shear stiffness.

Some polymers are crystallizable; that is, they form tightly bonded, highly regular three-dimensional arrays of themselves (Figure 5.6). Even if this happens in small patches, the

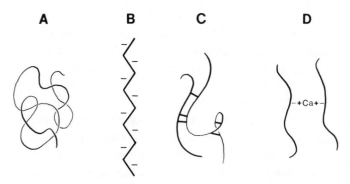

Figure 5.5 Control of molecular size and shape by charges. (A) Uncharged, coiled, randomly thrashing molecule. (B) Negative charges repel each other and cause it to straighten out. (C) Charged groups lead to short cross-links. (D) Divalent cations such as calcium and magnesium can form cross-links between negative charges. These links may be broken and reformed according to the calcium demands of the immediate environment. Drawings by L. Croner.

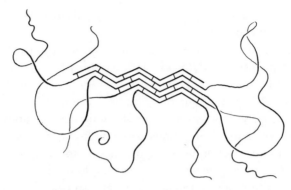

Figure 5.6 Crystalline regions of polymer mixtures occur when regions of identical structure become oriented parallel to each other and cross-link. Drawing by L. Croner.

inclusion of small, stiff regions in a matrix stiffens the matrix. It is common for crystalline polymers to form fibrous structures (cotton, collagen, chitin, and silk fibers are examples) that contain so many aligned molecular strands that the array is visible to the unaided eye. These fibrous materials are among the stiffest polymeric materials in nature.

Calcium and magnesium not only can link polymers directly but also can form crystals of carbonate or phosphate salts (Figure 5.7) that may bond to polymers or act as fillers in making up bone, coral skeleton, sea urchin stereom, crustacean cuticle, mollusk shell, coralline algae, and stromatolites. These are the stiffest materials in the living world.

It should be clear by now that we have the ingredients to make every structural material known to us from plants and animals, past or present. These materials did not all evolve at once or only in the sequence implied here. But what is evident is that a few polymers and a few other chemical compounds have been combined by ancient and modern organisms to make a variety of materials that have different mechanical properties. That increasing shear modulus may have been a trend at times in evolution has been postulated (Figure 5.8) for the mucoid secretions and cuticles of spiralian worms (Turbellaria, Archiannelida, Annelida) by Rieger and Rieger (1976), for the molluskan shells by Stasek (1972), and for plants and animals in general by Gordon (1978).

Compared with fluids, organisms are stiff, but compared with the inorganic, solid materials in nature, organisms are compliant. Solid material of low stiffness is one of the most important things that organisms have brought to Earth. We associate the height of our advanced state of progress in science and engineering with the development of metals and ceramics that allow us to make the biggest, stiffest, and fastest solid things in which to live, work, and travel to the moon. We should not forget that vital early steps in our material technology involved the use of soft, flexible materials from nature in the fabrication of clothes and cordage for everything from fishing line to suspension bridges. Only since the 1940s have we tackled the task of copying and trying to improve on nature's softer materials in making synthetic fibers, polymeric heart valves, and Teflon hip joints.

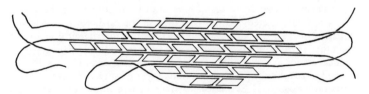

Figure 5.7 In bone and shell, mineral crystals (*parallelograms*) can form on charged sites along polymers. This deposit stiffens the resulting material. Drawing by L. Croner.

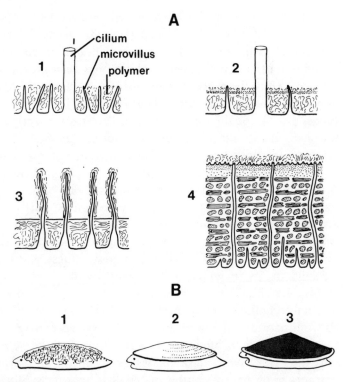

Figure 5.8 (A) Some protective polymeric "cuticles" in animals drawn from transmission electron micrographs. Polymer layer on an epidermal cell of (1) *Nemertoderma* and (2) *Retronectes,* both turbellarians; (3) *Protodrilus,* an archiannelid, and (4) *Lumbricus,* the annelid earthworm. These are arranged here in order of increasing complexity. The large striated and dotted forms in the earthworm cuticle are collagen fibers in layers. Nothing is known of the chemical nature of the other textured parts shown in any of these cuticles. (B) Hypothetical stages in the evolution of the molluskan shell. (1) A soft-bodied, wormlike animal with a thick layer of sticky, mucous polymers on its back. (2) Evolutionary stage in which the polymer coat is cross-linked. (3) A mollusk with a fully calcified polymeric shell. Drawings by L. Croner.

FEATURE 3: PARALLEL
ORIENTATION OF FIBERS

All things, including polymers, are oriented in space. Polymers may have random orientation with respect to each other (Figure 5.9A); they may have parallel orientation (Figure 5.9B), as cellulose does in a crystalline microfibril in a plant cell wall; or they may have any degree of preferred orientation (Figure 5.9C,D).

Preferred orientation of macromolecules is exceedingly common both inside and outside cells in plants and animals. Molecules may be aligned by the self-assembly process that takes place in specific sites at the cell surface. Alignment from a slurry of randomly scrambled molecules may be caused by the distributions of charges and nonpolar sites on the molecules.

Attribute: Anisotropy Fibers or long, straight-chain polymers confer different values of mechanical properties in different directions in materials according to the degree and direction of their preferred orientation. Either a random or a layered "plywood" array of stiff fibers oriented in the plane of a membrane will cause the membrane to have the tensile stress–strain curve shown in Figures 3.7 and 3.11C. This J-

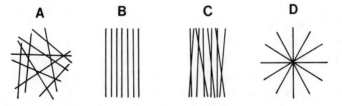

Figure 5.9 Orientation of long, thin elements. (A) Random orientation. (B) Parallel orientation. (C) Preferred orientation. (D) Radial or spherulitic orientation. Drawings by L. Croner.

shaped curve is a fundamental characteristic of most soft tissues (Gordon, 1978; Yamada and Evans, 1970). The curve shows high compliance in the initial part of the curve—a small stress creates an enormous deformation—because the polymers are being uncoiled and reoriented rather than being directly stretched parallel to their long axes. This uncoiling stage accounts for the initial softness and stretchiness of compliant polymeric biomaterials.

Equally important is the increase in stiffness indicated by the increasing slope of the curve as stress is increased. The stiffness (slope of the curve) of the material rises until it has nearly the same stiffness as the fibers themselves. As deformation increases, the fiber angle with respect to the stress axis decreases and the material becomes stiffer. This is an important structural safety factor.

We associate movement of cells and cell organelles with parallel polymers. The constriction of the cell membrane of dividing sea urchin eggs is accompanied by an increasingly parallel array of actin filaments immediately underlying the constriction (Schroeder, 1973). The contraction of muscle and the motion of flagella are both mediated by shearing between molecules in parallel polymeric arrays.

Perhaps most important here is the observation that fibroblasts, the human connective tissue cells that are the most commonly cultured cells in cell research, divide and move along ridges or grooves on the substratum (Weiss, 1955; Overton, 1979). Masses of fibroblasts—not unlike the blob of cells we are considering—grow preferentially along such irregularities on the substratum. When grown on thin sheets of silicon rubber, fibroblasts exert tractional forces on the substratum (Harris et al., 1981). The moving cell pinches the membrane immediately underneath itself, thus applying compressive stresses to it (Figure 5.10A). This wrinkles the membrane underneath the cell. Beyond the cell, the same pinching creates tensile stresses, thus stretching the rubber membrane and forming ridges that extend radially away from the cell. When two blobs of these cells are grown on collagen gels whose collagen is initially randomly oriented, the blobs pull on the gel material between them and cause the collagen in the gel to become oriented (Figure 5.10B). The cells of the blobs finally touch one another, attach, and then contract. In this way, they form a bridge of highly oriented cells and gel between the two blobs (Figure 5.10C).

The activity and growth of cells along these exaggerated stress lines is increased until a bridge of cells and collagen fibers joins the two shapeless blobs into one cylindrical one. In other words, the behavior of these cells is governed by the parallel orientation of the macromolecules of their intracellular cytoskeleton. The cells cause the parallel orientation of extracellular polymers by pulling on them and, having oriented them, the cells then move and grow along the structural features they have made in the substratum. And as they move, they form more collagen fibers that are oriented to their direction of movement.

Should cells in our primordial blob have created parallel orientation of their extracellular polymer on the surface of the blob, they too may have responded to the oriented polymer by increased growth in the direction of orientation: the blob became longer in that direction and became a cylinder. No matter what causes one or a group of cells to initiate preferred orientation of extracellular polymers, cells are capable of responding to the structured substratum (in this case a coating of the blob made of oriented extracellular polymers) by orienting their locomotor and growth activities preferentially parallel to the orientation of molecules in the substratum. Thus, there seems to be a causal relationship between the

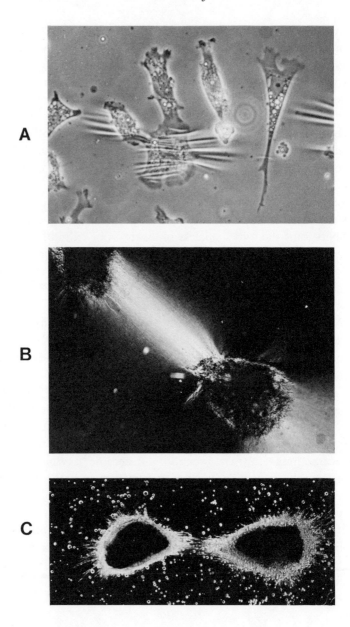

structural feature of preferred molecular orientation and the creation of a cylindrical body from a shapeless one via anisotropy, a functional property.

FEATURE 4: CYLINDRICAL BODY FORM

Well, here we are! The actual form that is most prevalent in multicellular plants and animals seems to have arisen inevitably from the secretion of a sticky extracellular polymer, its subsequent interaction with other polymers, and their preferred orientation.

Attribute: Polarity Polarity leads to bilaterality, streamlining, and locomotion. Even though a spherical body will greet the environment in the same way from all directions, the environment itself is basically anisotropic. Gravity pulls only downward. Light comes mainly from above. Wind and waves come from the side and often have a preferred direction, at least while the wind is blowing or the tide is flowing. If locomotion is to be at all efficient, it becomes desirable for motile organisms to reduce drag by minimizing the surface area that faces upstream. A fixed cylindrical body has the potential to resist gravity by having a design with a high flexural stiffness while retaining the ability to reach out. Motile bodies can point into the wind or current. Sense or-

---

Figure 5.10 Photomicrographs of chick heart fibroblast cells in culture. (A) Cell traction causes compression wrinkles in the rubber membrane on which they are grown. The long cell at the right is about 0.1 mm long. (B) Cells form a blob (about 1 mm across), move outward from the blob, creating traction on the collagen gel they are grown on. Neighboring blobs pull hard enough to orient the collagen molecules in the gel, as shown by the bright streak in polarized light. (C) Further cell growth and proliferation along the lines of tensile force and oriented collagen gel cause the two blobs to become one cylindrical body.

gans, gills, and mouth(s) collected on the front end of motile bodies and accompanied inside by the neural circuitry necessary to run this equipment confer polarity and a head end to the body. A cylindrical animal with its head oriented upstream may prefer to have one side down. If it matters which way is up and statocysts arise that can sense the direction of gravity, or if there are other structures that may lie asymmetrically in the body with respect to gravity, the top and bottom sides are defined. These are necessarily accompanied by right and left sides. So a cylindrical creature that meets environmental factors at one end is polar, and one that faces upstream and prefers gravity to come always from the same morphological direction, or prefers one side against the ground, is bilaterally symmetrical. In short, cylindrical bodies are more streamlined than blobs and spheres are.

ALTERNATIVE FORMS

Consider alternatives to the cylindrical form. What if most or all organisms were in fact spherical (solidly three-dimensional) or platelike (two-dimensional) in shape?

What do organisms *do* with their bodies? Organisms take in nutrients and light from the environment in order to grow and reproduce. Reproduction is the zenith in the life of an organism. Growth is a means to ensure collection of air- or waterborne gametes for reproduction. Growth also is necessary if one is to reach out and compete successfully for oxygen, nutrients, sunlight, or gametes in the wind; but achieving height to outreach a neighbor by growing spherically is expensive. Absorbing nutrients, catching prey, and absorbing sunlight and oxygen happen at an organism's surface. As a spherical body grows, its surface increases as the square of linear dimension, but its volume of tissue to be fed increases as the cube of linear dimension. It would be highly imprac-

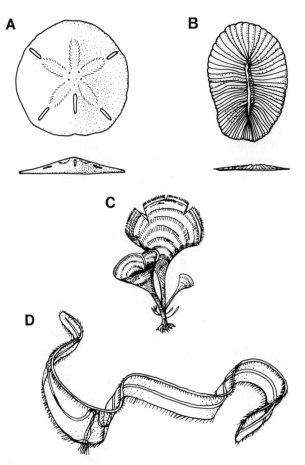

Figure 5.11   Platelike (two-dimensional) organisms. (A) Sand dollar (*Mellita;* Echinodermata), top and rear views. (B) A particularly ancient fossil animal, *Dickinsonia;* top and edge views. (C) A tropical seaweed, *Padina* (Phaeophyta). (D) Venus' girdle, *Folia;* a transparent, gelatinous ctenophore. Drawings by L. Croner and L. Srba.

tical to use this growth form if it had to compete with a more cost-effective one.

Being plate-shaped makes more sense than being spherical, because thin, leaflike structures are virtually only area. There are platelike and leaflike organisms: sand dollars are very flat, disk-shaped animals, and plants such as the alga *Padina* (Figure 5.11) are exactly that—just two layers of cells, and no more. One could imagine a world of leaflike organisms. Perhaps the reason our world is not like that is that another shape, the cylinder, outcompeted (shaded) the plate shapes: it may have had more advantageous design features than just the ability to produce nutrient-absorbing surface. It could certainly have swum, run, and flown better.

The cylindrical body form is prerequisite for many of the most characteristic features of plants and animals: their branching forms disperse their leaves, roots, mouths, and gamete exchangers into positions where they can compete for space, light, food, water, gas exchange, and incoming gametes. The abilities of animals to swim, crawl, run, jump, and fly appear also to be permitted by the cylindrical shape of their bodies.

6
PLANTS AND
ANIMALS TODAY

I have made a general statement that plants and animals
are cylindrical. This statement leads to the inference that the
mechanical factors of the environment have been important
determinants of organismic form. And it is compatible with
the theories of the stiffness of materials and the shapes of
load-bearing structures and streamlined bodies. Cylindrical
bodies also fit the theory that heredity, competition, and natural selection among organisms operating in wind and waves
in the earth's gravitational field have produced the plant and
animal forms we see today.

Does the existence of noncylindrical organisms invalidate
the concept that mechanics is a major determinant of form?
Increase in size and complexity are two major trends in the
evolution of modern plants and animals. Can the mechanical
design principles of cylinders also provide an explanation for
some changes in architecture from small, simple organisms to
larger, more complex ones? How well does the theoretical,
mechanical explanation of form fit what we know about
form and function in real cylindrical organisms? These questions will be answered in the following sections.

THE SHAPES OF ORGANISMS

Three-Dimensional Lumps

Most organisms are cylinders, but some are two-dimensional fans and plates and some are more-or-less shapeless, three-dimensional lumps. And organisms such as sponges, soft and stony corals, bryozoans, and calcareous Red Algae can exhibit platelike, cylindrical, and lumplike shapes within the same colony. Although the lump body form could have preceded cylindrical forms in the early evolution of multicellular life, the blobs we see today are certainly not all primitive forms. Moundlike and hemispherical sponges and stromatolites may in fact be in evolutionary lines that never had another shape, but colonial corals and bryozoans surely started as solitary cylindrical polyps and later evolved sheetlike, lumplike, and branched colonial forms.

Spherical organisms meet the flowing wind and water and the radiating light and heat of the environment with the same profile in all directions. Not being streamlined, those that are not attached to the substratum will tumble, as *Volvox* does in midwater and as various calcareous Red Algae and a few corals do in the wave-swept surge channels on the tops of coral reefs. Planktonic medusae, ctenophores (Figure 5.11), and larval forms of many worms, mollusks, and echinoderms have lumplike or nearly spherical shapes. They make much less headway by swimming than they do by being tumbled about by turbulence and by riding ocean currents. They feed on smaller plankton by creating local currents with tiny, hairlike cell organelles, the cilia. Such organisms should not be expected to be streamlined (cylindrical), because they depend on drag forces for their dispersal.

Lumplike corals, stromatolites, and reef-building Red Algae are alive only at their surface. The blob-shaped body is created by the addition of layers; growth starts at the point of

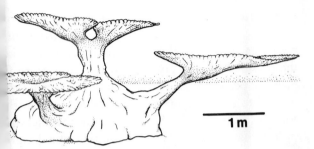

A colony of the table coral *Acropora reticulata*, as it
l lagoons. Drawing by L. Croner.

ts (for example, the Rhyniophyta) were cylin-
and that leaves of some living plants, especially
are slender cylinders (needles).

, appendages are either respiratory or mechanical
Respiratory gas exchange, like photosynthesis,
surface area, and gills are filamentous or plate-
omotor organs that create thrust or lift against
also depend on large areas: fins and wings are fan-

species of plants and animals, noncylindrical
indeed, advantageous. They are the basis for
ures and functions in locomotion, feeding, and
. They occur more often as appendages (leaves,
and hands) than they do as entire bodies (sea-
), and in most of those cases they are supported
l elements or bodies.

Cylinders

ia, fungi, and algae make simple, cylindrical
parallel or interwoven filaments, each of which
of cells (Figure 6.3). For the most part, the cells
ts are cylindrical; and each one is encased in a

---

attachment of the propagating larva or spore and moves outward to form a small plate. This plate then continues to increase in thickness at about the same rate as it increases in
diameter. In fact, the cylindrical form of plants and animals is
most often generated by an initially blob-shaped body whose
axial growth rate greatly exceeds the rate of growth in diameter.

Sea urchins, turnips, barrel cacti, and their relatives each
have a blob-shaped component that dominates their shape. A
turnip is a locally thickened root, but its stems, petioles, and
secondary roots are cylindrical. Immediately following their
metamorphosis, sea urchins grow as orbs; but they reach out
and move around on cylindrical appendages: stiff spines and
flexible, hydrostatic tube feet (Figure 6.1). It is more important for barrel cacti to store water than it is to be tall, thin, and
spreading. Sunlight is abundant, water is scarce, and a spherical shape is the simplest shaped jug to store it in. Turnips and

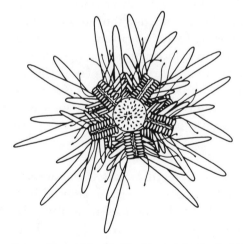

Figure 6.1 Sea urchin; this organism has a subspherical body and
cylindrical spines and tube feet. Drawing by L. Srba.

onions store food in inflated parts of their otherwise cylindrical roots and stems; and the globular sea urchin body contains a large, bizarre, five-jawed feeding apparatus and a clutch of other internal organs. All of these organisms use tubular structures to distribute food throughout their bulky forms. Clearly the evolutionary and developmental mechanisms producing blobs are as diverse as the functional approaches of those organisms are to life in their habitats: sponges filter bacteria from the passing stream, sea squirts (phylum Urochordata) filter phytoplankton; corals and bryozoans catch zooplankton; and algae and most reef-building corals make food by photosynthesis.

Noncylindrical forms exist in all animal phyla and plant divisions; and, although noncylindrical bodies do not have the same mechanical advantages that cylindrical ones have, they presumably have nonmechanical capabilities that outweigh any mechanical disadvantages their shapes may have. After all, the fitness of a species is the result of all its advantages and disadvantages. Every complex design is a compromise because of the conflicting functions that must be served. Moreover, noncylindrical organisms almost always have cylindrical supportive, locomotor, and circulatory body parts. The occurrence of cylindrical parts of noncylindrical bodies is evidence of the importance of local stiffness and the movement of body fluids within plants and animals just as cylindrical body form is important in the environment.

Fans and Plates

Plate-shaped organisms that attach to the substratum and extend into the flowing water are streamlined if their plane is parallel to the flow: this posture minimizes drag forces by minimizing the surface area that is projected into the flow. Because incident sunlight comes mostly from above and

evenly from all sides, s̲
sonable way for plants ar
high drag. But plate-shap
the passing stream may
plane is perpendicular t̲
normal to the flow max̲
tear these animals off th̲
mise between rigidity ar̲
port system is of great i̲

In the face of waves,
broken or swept away. ̲
the second is to be flexi̲
more streamlined shape
to lie parallel with the
high drag forces. Bei̲
being stiff and strong
achieves a flattened forr̲
to present a stiff wall p̲
and bryozoans build r̲
flow. Either these anin̲
and brings adequate fo̲
in sheltered places, an̲
pumping. The great t̲
of Pacific atoll reefs (
fragile. They live in h̲
movement is purely h̲
and offers no danger

Broad, flat leaves a̲
are stiff enough to ach̲
per unit mass than c̲
linders. Although thi̲
leaves with large su̲
noteworthy that the ̲

Figure 6.2
grows in co̲

restrial pla̲
drical stem
the conifer̲

In anima̲
in function̲
depends on
shaped. Lo
water or ai̲
shaped.

For som̲
shapes are,
specific pos
reproductio
fins, wings,
weeds, cora
on cylindric

Some bacte
thalli of man
is a long cha̲
in the filam̲

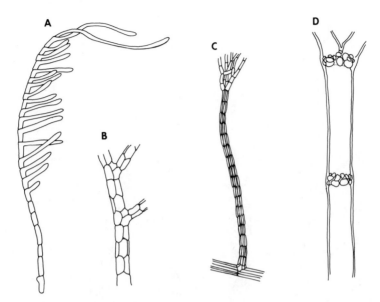

Figure 6.3 Algal filaments of one or a few cells in diameter. (A) _Cladophoropsis;_ (B) _Falkenbergia;_ (C) _Herposiphonia;_ (D) _Ceramium._ Drawings by L. Srba.

wall that is reinforced by helically wound filaments of cellulose or chitin. Of all the lumpy, platelike, and filamentous forms that may have appeared in the earliest stages of evolution, it is the filamentous ones that have given rise to most modern plants: the basic, original multicellular plant was a filament.

It is reasonable to suppose that the first motile, multicellular animal was a tiny worm living on or in the seafloor. Today there are species of tiny worms (< 1 mm in diameter) in most of the two dozen phyla treated in textbooks of invertebrate zoology (for example, Barnes, 1980) (Figure 6.4).

A worm is essentially a sock full of meat. In most worms, the sock is a fabric of crossed-helical collagen fibers, and the

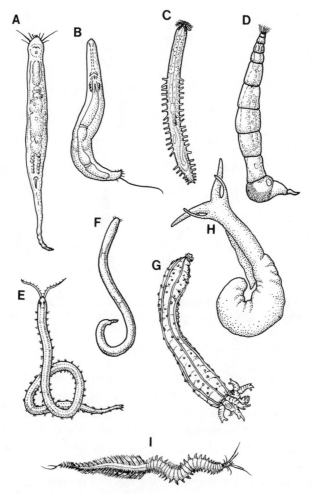

Figure 6.4 Wormlike animals from nine phyla. (A) *Gnathostomula* (Gnathostomulida); (B) *Myozonaria* (Platyhelminthes); (C) *Macrodasys* (Gastrotricha); (D) *Albertia* (Rotifera); (E) *Protodrilus* (Archiannelida); (F) *Enoplus* (Nematoda); (G) *Labidoplax* (Echinodermata); (H) *Microhedyle* (Mollusca); (I) *Heteronereis* (Annelida). Drawings by L. Croner.

meat is a system of muscle cells whose hydrostatic interaction with the sock produces locomotor movements. Active motion in wormlike bodies of all animals is produced by bending the body or by waving fins and wings, but it is also a function of arms, legs, fingers, tongues, trunks, tentacles, tube feet, and tails; and these appendages are all cylindrical. They accommodate the tensile and compressive stresses and strains experienced during bending along the limb's axis, and many of them can do so while bending in more than one direction.

INCREASING SIZE AND DISTRIBUTION OF MATERIALS

If early plants were filaments of cells and early animals were tiny worms, two obvious subsequent evolutionary trends in morphology of both plants and animals must have been increase in body size and increase in complexity of bodies and their components. As a cylindrical skeletal element grows, its strength and stiffness increase with its cross-sectional area, which is directly proportional to the square of its radius. Meanwhile, the body mass to be supported is increasing at a rate nearer to the cube of its radius. Therefore, a small organism cannot grow into a large one without changing the proportions or material stiffness of its supportive elements. In fact, we find that large plants and animals have a greater proportion of their body mass invested in rigid supportive structure than do small ones. Stiffness against gravity or wind and water flow as well as stiffness in aid of locomotion are dependent on rigid material, the shapes of skeletal elements, and their placement in the body. Even though organisms of all sizes make rigid materials, the wood, bone, and arthropod cuticular materials and their positions in the largest plants and animals make up some of the most sophisticated structures on earth.

When it comes to stiffening the entire wormlike body, we recognize two approaches alluded to by the title of this book: the body can be supported by a stiff endoskeletal axis or by an exoskeleton that places the stiff support at the circumference of the cylindrical body.

Axis

The backbone of vertebrates is a column of calcified lumps, the vertebrae, which are usually interspersed with soft, viscoelastic, intervertebral disks or capsules. The backbone is formed, during development, from a flexible, non-mineralized, collagen-wound hydrostatic rod called the notochord, which lies between the gut and the tubular central nervous system (spinal cord). That's what a backbone is. This is what it does: it resists changes in the length of the body while allowing the body to bend.

The most familiar backbones are those of the subphylum Vertebrata. Cartilaginous and bony vertebrae evolved as hard parts along the notochord, which can still be found in adult amphioxus, lampreys, and sturgeons. Notochords also appear in the embryonic stages of sea squirts and salps and most vertebrates, but not in the adults of these animals. Animals with soft backbones must have evolved from wormlike animals without recognizable backbones. It is apparent, however, that animals of many invertebrate phyla have structures that resist changes in body length while permitting bending. Some of these structures in flatworms and gastrotrichs are like notochords; and others, particularly those in the arms of brittle stars and sea lilies, bear a resemblance to our own backbones: each is a series of calcified ossicles joined together by flexible, collagenous ligaments. Furthermore, the notochord-like structures in all worms and vertebrates (but not in brittle stars and sea lilies) arise during embryonic development as part of or as extensions of the dorsal wall of the gut.

A true notochord in a chordate body is literally a worm within a worm. In the tadpole larvae of some tunicates and amphioxus, the notochord is a column of vacuolated cells. In some free-swimming adult tunicates such as *Oikopleura,* the notochord is a hollow, fluid-filled collagenous tube; and the jawless lampreys and hagfishes have a collagenous notochord full of a viscous gel. In all these animals, collagen fibers wrap the hydrostatic notochord in a crossed-helical array. Like invertebrate worms, both the notochord and its owner are membrane-wrapped hydrostats bent by sheets of longitudinal and circumferential muscles that apply distributed loads to these membranes.

The appearance of rigid material in the skeleton enabled the subsequent development or evolution of tendons. When the embryonic notochord of vertebrates becomes further stiffened by ossification, it does so in a series of vertebral bodies along the notochord. Each lump of bone or cartilage provides a solid attachment site where muscles, organized into large, discrete units, can apply their large forces as point loads to the backbone via tendons.

The backbone is put into longitudinal compression by every contraction of longitudinal muscle, and the soft structures between the vertebrae are required to do all the bending while resisting large changes in body length. Each intervertebral unit consists of a ring, the fibrous annulosus, of overlapping collagen fibers around a fluid- or gel-filled center. Collagen fibers are cross-helically wound around the short, cylindrical fibrous annulosus.

The fact that we have named the supportive rod in one phylum the backbone has kept us from appreciating the wide distribution of axially stiff, flexible structures supporting wormlike bodies throughout the animal kingdom. In addition, many arthropods are built to resist changes in length and to allow flexure. The backbone-like structure in ar-

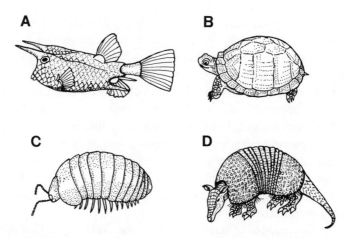

Figure 6.5 Cow fish (A) and box turtle (B) have peripheral rigid skeletal structures. Pill bugs (Crustacea; C) and armadillos (D) are flexible because their crusty coverings are jointed. Drawings by L. DeLeon.

thropods is not an axial rod but a jointed exoskeleton arising from the stiffening of the body wall (the circumference).

In the evolutionary radiation of paired appendages and different modes of locomotion, situations arose in which body flexibility was either a liability or simply not significantly advantageous. Thus, in the bodies of birds, the vertebrae, ribs, and breastbone have fused into a rigid tube. With the locomotor mechanism out in the wings, the best body is a rigid, streamlined rocket. Cow fishes and box turtles (Figure 6.5) enjoy the protection of a boxlike shell. And in puffer fishes (Figure 2.1E), the ability to swallow thrice the body volume of water, and thus achieve much greater size and an armory of erected bony spines, has been more advantageous than body flexibility. In all three cases, the vertebral columns of these animals are fused (Figure 6.6).

Some of the greatest advances in the design of rigid support elements among terrestrial organisms are the beamlike axial supports of branches, arms, legs, and wings. Each of the great lower branches of the live oak, *Quercus virginiana,* may extend 30 m horizontally (Figure 6.7). This species is obviously preadapted to shade parking lots. It is fascinating to sight along one of these branches during a wind storm. The branch's shape is a gentle helix, and as the wind buffets it up and down and from side to side, the spring in the helix "works" by visibly shortening and lengthening. This could well be an effective way to distribute bending stresses in the branch so as to reduce maximum strains to levels that would avoid breakage.

The long bones that are the axial support of the legs of elephants, giraffes, and other large quadrupeds are hollow beams that accommodate the compressive and bending stresses placed on them. These bones have expanded ends that distribute the compressive forces that occur on the im-

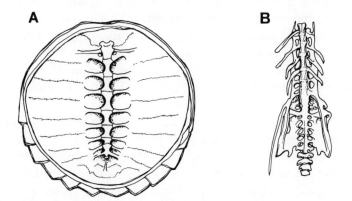

Figure 6.6 Backbones of turtles (A, alligator snapping turtle) and birds (B, brown pelican) are stiff because the vertebrae are fused. Drawings by L. Croner and L. Srba.

Figure 6.7 Live oak tree branches (*Quercus virginiana*) are often longer than the tree is tall. Drawing by G. Minnich.

pact of each running step. The specific rounded shape of the ends of long bones also allows for the transmission of compressive forces across the flexible joint, even while the joint is being bent over a wide range of angles.

Bones that are the axial support of vertebrate limbs have the same function that backbones do. The long, slender fingers of bats and extinct flying reptiles (Figure 6.8) are graphic examples. They support the tightly stretched, tensile wing membranes by their own compression: they resist change in length while permitting controlled flexure at the joints.

Other animals and plants have bodies stiffened by rigid material placed externally to form exoskeletons. The famous calcified husks of lobsters, millipedes, and beetles come to mind, as well as the bony shells of cow fishes, turtles, armadillos, and extinct glyptodonts (Figure 6.5). If one were to stiffen a worm, an obvious way would be to stiffen the soft, polymeric cuticle by cross-linking structural proteins and polysaccharides or by including rigid mineral crystals into the external, membranous cuticle. It is likely that the complex, fibrous, layered cuticles we see in large worms and arthropods evolved in just this way.

Because bone is produced by cell layers below the epidermis, the dermal bones that cover armadillos, gar fishes, and turtles are seldom called exoskeletons, but functionally, that is what they are. Similarly, the arrays of calcified ossicles produced in the mesoglea of soft corals and in the dermis of most echinoderms, some worms, and fishes (Figure 6.9) serve as exoskeletal materials. If one backs away from the microview and looks at the placement of these rigid structures in the body of these animals, they are, mechanically speaking, exoskeletons. There are both soft and rigid exoskeletons in most phyla.

Although the scaly covering of most fishes and pangolins and the feathery coverings of bird bodies do not serve in the transmission of locomotor forces, they protect the bodies effectively from mechanical blows. Gordon (1978) has

Figure 6.8 Finger bones in bats' wings support the wing membrane as a mast supports the sail of a boat. Drawing by L. DeLeon.

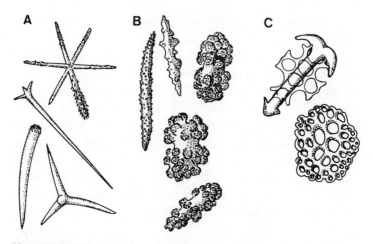

Figure 6.9 Microscopic, calcitic ossicles reinforce collagenous supportive tissues in the bodies of sponges (A), soft corals (B), and sea cucumbers (Echinodermata; C). Drawings by L. DeLeon.

pointed out the value of a layer of feathers in the absorption of energy of a blow. He notes that armorers employed by shoguns in medieval Japan made body armor of overlapping feathers or porcupine quills.

Even though gravitational force on aquatic organisms is alleviated by their buoyancy, large fishes, turtles, and porpoises that move rapidly through the water have rigid endoskeletal materials. Other organisms whose great size is buoyed up by the ocean's water either swim very slowly or not at all: medusae that drift and giant kelps that float have no rigid materials to support their bodies and maintain their form.

Terrestrial plant stems, thalli of intertidal seaweeds such as the sea palm, *Postelsia,* and colonies of soft corals flex with the wind and waves. The stiffer of these also resist changes in length. Unlike the multimaterial joints of echinoderms, arthropods, and vertebrates, these soft plants and animals flex with complete continuity of material—they are branched cylinders and they span the categories of axial and circumferential stiffening. Beam theory tells us that the most economical means of stiffening a cylinder is to place the stiffest material as far away from the axis as possible—at the circumference. Stony coral colonies, wilting plants, most large trees, and the elegantly designed bamboo depend on circumferential materials for most of their flexural stiffness, whereas flexible sea fans and seaweeds are built with their stiffest material on or near the body axis. Their flexibility is allowed by the low second moment of area created by the central placement of stiff material.

One of the most remarkable achievements of organismic evolution is the array of stiffness of bending cylinders: from the most flexible worms, seaweeds, and cats to the rigid sequoias, turtles, and corals. Some stiffen the axis and others stiffen the circumference.

EVOLUTION OF WHOLE-ANIMAL DESIGN

The earliest, tiniest organisms lacked specialized organs for respiration and circulation: they could depend on diffusion for the transport of dissolved substances to and from all parts of the body. With an increase in the thickness of the body came the food- and water-conducting systems of plants and the gills, hearts, blood vessels, lungs, and tracheal systems of animals. Each system also became more complex: whereas the gut of tiny worms today is frequently a simple tube consisting of a single layer of cells, various larger animals have tentacles, mouth, pharynx, esophagus, gizzard, crop, stomach, rumen, and small and large intestine. Presumably each of these specializations allows for some advantageous energetic efficiency.

A   B

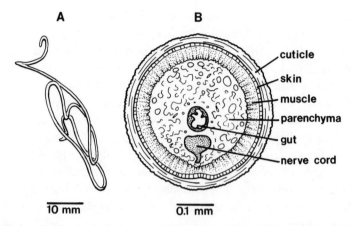

cuticle
skin
muscle
parenchyma
gut
nerve cord

10 mm   0.1 mm

Figure 6.10 A gordian worm (Nematomorpha). (A) It is long, thin and flexible. (B) Cross section showing the extensive spongy parenchyma between the body wall and the gut. Drawings by L. Croner.

Flatworms (phylum Platyhelminthes), ribbon worms (phylum Nemertea), and horsehair worms (phylum Nematomorpha) have no general body cavity other than the gut. An extensive, soft parenchymal tissue fills the body between the gut and the muscular, collagen-wrapped body wall (Figure 6.10). These animals live in or on solid or muddy substrata. They have only slow-acting smooth muscles, and it is likely that the viscosity of soft parenchyma contributes to the slowness of movement.

In the first thorough functional analysis of animal structure and phylogeny, Clark (1964) concluded that a secondary body cavity (the gut being the primary one) and a segmented body were essential features in the evolution of animals that lived in water or burrowed in the sediments of the seafloor. I will add to that the crossed-helical array of collagen fibers that controls shape changes in soft-bodied animals.

Spacious body cavities filled with fluid occur in some worms (nematodes, echiurans, sipunculans, and some annelids). The fluid in spacious cavities allows the instantaneous transmission of pressure created by muscles to all parts of the body. This can help a worm mobilize its entire locomotor system at once. Many worms with spacious body cavities are filter or deposit feeders that live sedentary lives in sediments, often in tubes of their own manufacture. Their locomotor activity consists of maintaining their position and posture at the mouth of their tube where food and oxygen are available and withdrawing quickly into the tube when threatened by a predator. Their muscles are often the helically striated ones whose speed of contraction is intermediate between those of smooth and cross-striated muscles. Their fast withdrawal into a tube is promoted by giant nerve fibers that carry the excitation more rapidly to all longitudinal muscle cells than do the smaller diameter neurons of flat, horsehair, and ribbon worms.

Body cavities, whether spacious or not, have other important functions. In some phyla, they are sites for the collection of urine. In all hollow animals, eggs are formed in the wall of the cavity and are released into the cavity. In some, fertilization may take place in the cavity; in others, eggs move through the cavity to a uterus for fertilization; still others release eggs to an external watery environment where fertilization takes place.

Even though the excretory and reproductive roles of body cavities have been known and much discussed for over a hundred years (see Libbie Hyman's multivolume treatise, *The Invertebrates,* 1940 to 1967, for an exhaustive review), their mechanical role as hydrostatic skeletons was not thoroughly put into an evolutionary context until Clark's (1964) provocative analysis. After citing the morphology and behavior of all phyla in turn, he concludes that the original function of spacious body cavities must have been the hydrostatic one that promotes posture and locomotion.

In other hydrostatic animals of all sizes, such as snails, caterpillars, and vertebrates, the body cavity is full of organs made of soft tissue: there is no spacious fluid-filled cavity. Each organ has its own covering, and adjacent organs can therefore be moved independently, except for their points of attachment where nerves and blood vessels come to them. The most sophisticated and complex animal designs are those in which appendages arise from the body wall. In some animals, such as insects, birds, elephants, and giant ocean sunfishes, the appendages are the locomotor organs, and the structural features of the body that promote undulation are sometimes reduced or lost altogether. Arthropods and chordates have fast-contracting cross-striated muscles and systems of skeletal levers and tendons that promote speedy movement, including flight.

ENVOI

The first and therefore the basic multicellular organisms may have been shapeless blobs that acquired a body axis and became soft, flexible, filamentous plants or wormlike animals. As plants became larger and more complex cylinders, their increased diameter gave them greater scope for the diverse design of the cross section of stems and the use of the second moment of area. Part of their increased complexity includes a range of flexibility brought about by various patterns of distribution of stiff materials across their cylindrical organs. Trees and giant bamboo plants that support photosynthetic organs and wind-dispersed seeds 50 m off the ground have achieved the same architectural status among plants that the Eiffel Tower has among human-designed structures. Most of their support comes from material placed at the circumference of their cylindrical stems.

Meanwhile the first soft, flexible animals evolved into larger worms with secondary body cavities and crossed-helical collagen arrays in their skins or cuticles. They became eligible for a wide range of second moments of area of their cylindrical bodies. Like plants, they branched. Their increasingly complex locomotor abilities involved mechanisms for faster and more forceful movements. Two general systems evolved that resist change in length while permitting flexibility: the axial backbone and the peripheral exoskeleton. Stiff materials in either of these sites allowed the evolution of great muscle masses to effect forceful or fast locomotion by applying point loads to skeletal levers via tendons. Greater loads were exploited by greater levers. Flying birds, bats, and insects have achieved the same level of architectural and functional sophistication that small, maneuverable aircraft have among our own designs.

Both animals and plants have evolved a variety of ways to maintain posture in gravity, wind, and waves by the judicious design of the axis and circumference of their cylindrical bodies.

BIBLIOGRAPHIC NOTE

REFERENCES

ILLUSTRATION SOURCES

INDEX

Bibliographic Note

Excellent books on the specific topics treated in this book abound. The gentlest and most delightful books in science and the humanities I have ever read are the introductions to natural and manmade materials and structures by J. E. Gordon (1968, 1978). Alexander (1971), Currey (1970), and Neville (1976) take the subject a step further in biological context without loss of readability. These brief, general accounts show the way to more detailed and sophisticated accounts.

Informative and thorough texts and reference works are becoming increasingly common. Kollman and Côté (1968) provide a broad coverage of the mechanics of wood, while Mark (1967) tells how the chemistry, morphology, and mechanics of wood cells are interrelated. Currey (1984) does the same for bone. Advanced accounts of the biomechanics of plants and animals are those of Alexander (1983), Vincent (1982), Vincent and Currey (1981), and Wainwright et al. (1976). Fung (1981) brings continuum mechanics to biology, and Vogel (1981) puts plants and animals into their mechanically rigorous environments of wind and waves. The first major account of animal evolution based on functional morphology is that of Clark (1964).

References

Alexander, R. McN. 1968. *Animal Mechanics*. London: Sidgwick and Jackson. 2nd ed., 1983. Oxford: Blackwell Scientific Publications.
—— 1971. *Size and Shape*. London: Edward Arnold.
Arber, A. 1954. *The Mind and the Eye*. Cambridge: Cambridge University Press. Reprint, 1985.
Barnes, R. D. 1980. *Invertebrate Zoology*, 4th ed. Philadelphia: Saunders.
Bonner, J. T., ed. *On Growth and Form*. Cambridge: Cambridge University Press.
Borelli, G. A. 1680. *De Motu Animalium*. Rome.
Bronowski, J. 1978. *The Origins of Knowledge and Imagination*. New Haven: Yale University Press.
Challis, D. A. 1969. An ecological account of the marine interstitial Opisthobranchia of the British Solomon Islands Protectorate. *Philos. Trans. R. Soc.* B 255:527–539.
Clark, R. B. 1964. *Dynamics in Metazoan Evolution*. Oxford: Clarendon Press.
Clark, R. B., and J. B. Cowey. 1958. Factors controlling the change of shape of some worms. *J. Exp. Biol.* 35:731–748.
Currey, J. D. 1970. *Animal Skeletons*. London: Edward Arnold.
—— 1984. *The Mechanical Adaptations of Bones*. Princeton: Princeton University Press.
Denny, M. W., and J. M. Gosline. 1980. Physical properties of the pedal mucus of a terrestrial slug. *J. Exp. Biol.* 88:375–393.
Eakin, R. M., and J. L. Brandenburger. 1974. Ultrastructural features of a gordian worm. *J. Ultrastr. Res.* 46:351–374.
Fane, L., and R. A. Smullin. 1984. *Frank Smullin—Sculpture and Drawings*. Duke University, Durham.
Fung, Y. C. 1981. *Biomechanics: Mechanical Properties of Living Tissues*. New York: Springer-Verlag.
Gordon, J. E. 1968. *The New Science of Strong Materials*. Harmondsworth, U. K.: Penguin Books.

———— 1978. *Structures or Why Things Don't Fall Down*. Harmondsworth, U. K.: Penguin Books.

Haloun, G. 1951. Legalist fragments, Part I Kuan Tsi 55 and related texts. *Asia Major* 2:85–120.

Harris, A. K., D. Stopak, and P. Wild. 1981. Fibroblast traction as a mechanism for collagen morphogenesis. *Nature* 290:249–251.

Hyman, L. H. 1940–1967. *The Invertebrata*, vols. 1–6. New York: McGraw-Hill.

Kier, W. M. 1983. Functional morphology of the musculature of squid arms and tentacles. *J. Morphol.* 172:179–192.

Kier, W. M., and K. K. Smith. 1985. Tongues, tentacles and trunks: the biomechanics of movement in muscular hydrostats. *Zool. J. Linn. Soc.* 83:307–324.

Koehl, M. A. R. 1982. Mechanical design in spicule-reinforced connective tissue. *J. Exp. Biol.* 98:239–268.

Kollman, F., and W. A. Côté, Jr. 1968. *Principles of Wood Science and Technology*, vol. 1, *Solid Wood*. New York: Springer-Verlag.

Leversee, G. J. 1976. Flow and feeding in fan-shaped colonies of the gorgonian coral *Leptogorgia virgulata*. *Biol Bull.* 151:344–356.

Mark, R. E. 1967. *Cell Wall Mechanics of Tracheids*. New Haven: Yale University Press.

Marks, R. W. 1960. *The Dymaxion World of Buckminster Fuller*. New York: Reinhold.

Myers, E. R., and V. C. Mow. 1983. Biomechanics of cartilage and its response to biomechanical stimuli. In *Cartilage*, vol. 1, ed. B. K. Hall, pp. 313–341. New York: Academic Press.

Neville, A. C. 1975. *Biology of the Arthropod Cuticle*. New York: Springer-Verlag.

———— 1976. *Animal Symmetry*. London: Edward Arnold.

Overton, J. 1979. Differential response of embryonic cells to culture on tissue matrices. *Tiss. Cell* 11:89–98.

Peterson, J. A., J. A. Benson, J. G. Morin, and M. M. Ngai. 1982. Scaling in tensile skeletons: structures with scale-independent length dimensions. *Science* 217:1267–1270.

Picken, L. E. R. 1960. *The Organization of Cells and Other Organisms*. Oxford: Clarendon Press.

Riedl, R. 1978. *Order in Living Organisms*. New York: John Wiley & Sons.

———— 1983. *Fauna und Flora des Mittelmeeres*. Hamburg: Verlag Paul Parey.

———— 1984. *Biology of Knowledge*. New York: John Wiley & Sons.

Rieger, R. M., and G. E. Rieger. 1976. Fine structure of the archiannelid cuticle. *Acta Zool.* (Stockholm) 57:53–68.

Rieger, R. M., and S. Tyler. 1974. A new glandular sensory organ in interstitial Macrostomida. *Mikrofauna des Meeresbodens* 42:137–175. Mainz: Akademie der Wissenschaften und der Literatur.

Russell, E. S. 1916. *Form and Function*. London: John Murray. Reprint, 1982. Chicago: University of Chicago Press.

Schmidt-Nielsen, K. 1972. *How Animals Work*. New York: Cambridge University Press.

Schroeder, T. E. 1973. Actin in dividing cells: contractile ring filaments bind heavy meromyosin. *Proc. Natl. Acad. Sci. USA* 70:1688–1692.

Smith, C. S. 1981. *A Search for Structure*. Cambridge, Mass.: MIT Press.

Stasek, C. R. 1972. The molluscan framework. *Chem. Zool.* 7:1–44.

Swedmark, B. 1971. A review of Gastropoda, Brachiopoda, and Echinodermata in marine meiobenthos. *Smithsonian Contrib. Zoology*, No. 76.

Taylor, W. R. 1960. *Marine Algae of the Eastern Tropical and Subtropical Coasts of the Americas*. Ann Arbor: University of Michigan Press.

Teragawa, C. K. 1986. Sponge dermal membrane morphology: histology of cell-mediated particle transport during skeletal growth. *J. Morph.* 190:335–347.

Thompson, D'A. 1917. *On Growth and Form*. Cambridge: Cambridge University Press.

Vincent, J. 1982. *Structural Biomaterials*. New York: John Wiley & Sons.

Vincent, J. F. V., and J. D. Currey, eds. 1981. *Mechanical Properties of Biological Materials*. Cambridge: Cambridge University Press.

Vogel, S. 1981. *Life in Moving Fluids*. Princeton: Princeton University Press.

Vogel, S., and S. A. Wainwright. 1969. *A Functional Bestiary*. Reading, Mass.: Addison-Wesley.

Vosburgh, F. 1977. Response to drag of the reef coral *Acropora reticulata*. *Proc. Third Int. Coral Reef Symp.*, pp. 477–482.

Wainwright, S. A. 1983. To bend a fish. In *Fish Biomechanics*, ed. P. Webb and D. Weihs. New York: Plenum Press.

Wainwright, S. A., W. D. Biggs, J. D. Currey, and J. M. Gosline. 1976. *Mechanical Design in Organisms*. London: Edward Arnold. Reprint, 1982. Princeton: Princeton University Press.

Wainwright, S. A., F. Vosburgh, and J. Hebrank. 1978. Shark skin: function in locomotion. *Science* 202:747–749.

Weiss, P. A. 1955. Nervous system. In *Analysis of Development,* ed. B. H. Willier, P. A. Weiss, and V. Hamburger, pp. 346–401. Philadelphia: Saunders.

Yamada, H., and F. G. Evans. 1970. *Strength of Biological Materials.* Baltimore: Williams and Wilkins.

Illustration Sources

1.2.   Yamada and Evans (1970), fig. 79. Copyright 1970 by the Williams & Wilkins Co., Baltimore.

1.4C.   Andrew Smith, *Echinoid Paleobiology* (London: George Allen & Unwin, 1984), fig. 3.38.

2.1A.   Peterson et al. (1982), fig. 1. Copyright 1982 by AAAS.

3.6.   Photomicrograph courtesy of Tatsuo Motokawa.

3.7.   Unpublished data courtesy of Lisa Orton.

3.9.   Wainwright et al. (1976). fig. 5.16.

3.10.   Koehl (1982), fig. 6.

4.2.   Kier and Smith (1985), fig. 5.

4.9.   Clark (1964), fig. 19C, and Clark and Cowey (1958), fig. 3.

5.8A.   Rieger and Rieger (1976), fig. 9.

5.8B.   Stasek (1972), fig. 1.

5.10A,B,C.   Photomicrographs courtesy of Albert K. Harris.

6.3A,B,C,D.   Taylor (1960), plate 1, fig. 2; plate 67, fig. 6; and plate 72, figs. 8 and 12.

6.4A,C,E,F,I.   Riedl (1983), figs. 70, 76, 78, 107, 139, 144, 219.

6.4B.   Rieger and Tyler (1974), fig. 3B. By permission of Akademie der Wissenschaften und der Literatur, Mainz.

6.4G.   Swedmark (1971), fig. 1F.

6.4H.   Challis (1969), fig. 181.

6.10A,B.   Eakin and Brandenburger (1974), fig. 1.

Index